HISTORY OF
SCIENCE · AND · TECHNOLOGY
REPRINT SERIES

Mind and Matter

Man's Changing Concepts of the Material World

Cecil J. Schneer

IOWA STATE UNIVERSITY PRESS • Ames

Cecil J. Schneer is Professor of Geology and the History of Science at the University of New Hampshire. His most recent publication is on the common symmetry of snowflakes, virus crystals, alloys, and the perfect solids of Leonardo and Kepler.

This edition published 1988 by Iowa State University Press, Ames, Iowa 50010
Text reprinted from the original without correction

Printed in the United States of America

History of Science and Technology Reprint Series
First printing, 1988

Library of Congress Cataloging-in-Publication Data

Schneer, Cecil J., 1923–
Mind and matter.

(History of science and technology reprint series)
Reprint. Originally published: New York: Grove Press, c1969.
Bibliography: p.
Includes index.
1. Chemistry—History. I. Title. II. Series.
QD11.S29 1988 540′.9 88-8915
ISBN 0-8138-0013-7

CONTENTS

PREFACE

To undertake a work of synthesis in an age of specialization requires some explanation. Modesty and the customs of the academic world require that the fields of learning, like the operations of home construction, be divided among experts. For the plumber to lay a brick, for the physician to relieve a toothache would be acts of impiety, fraught with hidden dangers as well as inviting instant retribution in the form of union penalties. Similarly, the historian of science approaches his subject with decent humility. For example, he might regard everything outside the decade 1720–1730 as beyond his ken. In this the historian of science only follows the example of his material. It would be a rash X-ray mineralogist of clays who would undertake a volume on the X-ray crystallography of the sulfides.

This volume, like its predecessor, *The Search for Order* (later titled *The Evolution of Physical Science*), grew out of my experience in conducting a course in physical science for students of the liberal arts. Like the householder ignorant of the niceties of union rules, who sees only that able-bodied mechanics are standing about at $3.50 an hour while water gushes into the foundation from an open valve, the student does not understand why it is that chemists speak only to chemists, and then with difficulty. Frequently the student assumes that if his question has been turned aside with the reply that the matter is too esoteric for undergraduate understanding, it is because the instructor does not know the answer. Yet there are real and compelling reasons for specialization. The growth of knowledge and skills is such that the single human mind cannot expect to function professionally in more than one restricted area, and therefore the necessity for synthesis is all the greater.

We live in a credulous world. The scientist or the teacher who remains silent with the excuse that the matter is out of his field of specialization is abandoning the contest to others, perhaps of lesser qualifications. For the teacher, at any rate, there can be no excuse for silence. If the schools themselves, in whose care society has placed the burden of transmitting the culture of the past and the responsibility for the culture of the future, are to restrict themselves to the narrowest professionalism, then where and how will our society be preserved? If our students are not to learn of the past and if their present is to be a series of deep and isolated pits of technique, then our future is bleak indeed. Already we see the fruits of the intellectual cynicism of a world of specialists when some men devote their lives to the breeding of new diseases, others to the large-scale destruction of landscape. Technology places a new leisure within our grasp, but we can employ it no more sensibly than to drive frantically down the endless concrete ribbons that separate one unbearable facet of existence from another. The work of our hands, the work of our heads is disposed of by others who, not being professionally trained, themselves uncreative and unproductive, are not troubled by refinements of moral feelings. Without emotion the "new men" who are really the old men dispose of the work of poets and riveters, of carpenters and mathematicians. They are the men for whom there is only one reality, only one goal: to rule others. Theirs is the passion of nationalism, their activity politics–the rule of man by man, the acquisition of power. As science becomes the means to wealth, but even more the means to force and dominance, the "new men" turn to science, or rather to its direction and control. A vast industrial-military-political complex is the common characteristic of the advanced nations today and the envy of the newly emerging states which can hardly wait until they too can subordinate all human values, all free ideas, to nationalist passions and the hatreds of modern times.

This new world thinks nothing of the destruction of whole populations if national interests seem to require it. A nation rends its own people in the temporary default of an external enemy. In this time the duty of students, of scientists, and of teachers is clear. It is the duty of students to master the techniques of the present so that they may take on the responsibility for a sane and generous future. It is the duty of scientists to struggle for the proper use of their achievements. It is the duty of teachers to transmit the heritage of civilization and to combat credulity, to the end that human values shall not disappear.

If the history of man as it is commonly written is the long and bloody account of the murder, robbery, and deceit by means of which man has subdued man–the successions of kings, of wars, of invasions, and of conquests–the history of science is the account of the intellectual development which has always in the past accompanied the events of political history, the submerged portion of the iceberg which has nevertheless supported all the rest. We think of Marie Curie performing the labors of Hercules to separate the radioactive fraction from tons of pitchblende; of Biot who wept when Pasteur showed him the optically active separation from racemate; of Priestley reviving his asphyxiated mouse with oxygen. The names of Galileo's opponents are forgotten, the rulers of his world lost in a confusion of battles and intrigues and centuries, out of which we pick unhesitatingly men like Dante, Leonardo, Pico della

Mirandola. We study history so that we shall be something more than children. We study the history of science so that we shall see in those examples of the human beings before us, in their trials, their misfortunes, and their triumphant genius, examples for our own time and our own lives. We study history for its own sake, for the sheer pleasure of tracing the development of those ideas, the lives of those men, which in their intricacy, subtlety, and grandeur are a reminder of all that is best in humanity and of the boundless potential of mankind.

This volume is an attempt to trace the development of our ideas on matter and the differences in the material world which are *chemistry,* from their beginnings in Stone Age technologies when the chemical and intellectual activities of man were invented, to the present and the probable mechanism of replication of the genes. Although I work in one of the sciences of materials, I am not a chemist. Although some of this book is original historical research, much of it synthesizes the work of the painstaking scholars of the history of science. The bibliography does not adequately express my indebtedness to the many historians and interpreters of science who have preceded me and whose works have guided me in these labors. I must especially acknowledge my dependence on J. R. Partington, Emeritus Professor of Chemistry at the University of London, R. J. Forbes, and the other contributors to the *Oxford History of Technology* for much of the information on the beginnings of materials technology; Andrew van Melsen for *From Atomos to Atom;* J. M. Stillman for his analysis of alchemy; Joseph Needham for opening the world of Chinese science; and H. M. Leicester and H. S. Klickstein for their collection of sources. I am also grateful to Professor Albert Daggett, of the University of New Hampshire Chemistry Department, for reading the manuscript.

This book would never have seen the light without the encouragement and loyal support of my wife, Mary Schneer, and my children, Jean and David, whose patience I gratefully acknowledge.

With so many excellent histories of chemistry available or in preparation, with the several clear and accurate popularizations of science by Isaac Asimov and George Gamow, any further efforts may seem superfluous, but I believe that the present volume is not a repetition in mood or content. Each generation sees its history in a different light. Each student of science, each research scientist reads the past in the light of his own experience. In the same way, the teacher transforms his material, seeing it in the light of the requirements of teaching, transformed through its interaction with the minds of students. The aim of this book has been to try to introduce the principal ideas of chemistry through the history of their evolution. It is my hope that the material here presented will transcend the limitations of its presentation to the end of a wider dissemination of a knowledge of the nature of chemistry in particular and science in general. As war is too important to be left to the generals, so science is too serious a matter to be left to the care of specialists.

Milan, 1969

PREFACE TO THE REPRINT

The reprinting of *Mind and Matter* 18 years after its first appearance, is a source of particular gratification. The book had its inception in the wave of concern over science education that followed World War II. The unexpected surrender of Japan in the face of the application of an abstruse branch of theoretical science raised at once both our hopes for a generous future and the deepest anxieties of those whose participation in the nuclear bombings had alerted them to the new dangers that threaten the earth. But I was unaware of most of this in 1946, and it seemed to me then that all things were possible for science and an educated society. With a generation whose lives had been interrupted by war, I resumed my education in science, and from the beginning, eked out my veteran's stipend by part-time teaching. My interest in the history of science developed out of my joint efforts at simultaneously learning and teaching. Given the opportunity at Cornell to study the history of science with Henry Guerlac while taking my degree in geology, I began a dual career in science and the history of science.

In my last year as a graduate student, Henry Guerlac put me on the program at the annual meeting of the History of Science Society. We could all fit around one large table for the annual banquet and, before we ate, we introduced ourselves in turn. That year Dorothy Stimson was the president. I had already read and cited some of her work, so that it never occurred to me that the Society was antiwomen. It was also my first meeting with C. Doris Hellman who was then teaching introductory science and was to enthusiastically embrace my books as they appeared. I. Bernard Cohen, the editor of *Isis,* the Society's journal, invited me to submit my paper for publication and, after I did, gently but firmly forced me to put it into shape. Carl Boyer, the mathematician-historian, who later on was to be my colleague on the Dictionary of Scientific Biography Board, sent me a book on crystallography to review for *Scripta Mathematica.* This ready acceptance was immensely flattering and, more than that, reflected a cordiality that contrasted with the harsh competition that mars so much of academia.

My generation of students were uncritical admirers of science and technology. As we viewed the historical era that molded our *weltan-*

schauung, we had passed from Lindy's transatlantic flight to the eradication of smallpox. As a child, I had put on headphones and heard my father's voice resonating mechanically all the way from station KDKA in Pittsburgh. My childhood coincided with the childhood of the electronic age. That early broadcast dealt, in all innocence, with public health. The symbols of our wartime youth were the radar-coupled mechanical computer that directed antiaircraft batteries and the sulfa drugs and, later, penicillin that destroyed infection. They were only stages on the way to the ultimate antibiotic–the bombs over Hiroshima and Nagasaki. In mindless admiration, we did not distinguish between a Pasteur overcoming the ancient diseases of mankind and the managerial entrepreneurs putting together their teams and squads, the individual units of which are not people but industries. The Manhattan Project has become the archetype of a hyperscience, inextricably bound to government, politics, and industry, but our generation never considered the implications of this. All doubts seemed swept aside by Whitehead's fierce imperative: "Where attainable knowledge could have changed the issue, ignorance has the guilt of vice."

In that postwar springtime, I learned from James Bryant Conant, George Sarton, and C. P. Snow the Jeffersonian faith that science and learning were good in themselves and that the terrible potential for their misuse could be overcome by the widespread diffusion of their understanding, best promoted by the conjunction of science with history. How fortunate for me that I could spend my life in a cause in which I believed and one which I could so heartily enjoy. The pursuit of science, Robert Hooke wrote in 1662, "hath a pleasure in it like that of the senses themselves." The study of history, wrote Halifax, is like wrestling with a handsome woman. For nearly four decades now, I have never doubted it. My early lectures on science and history were first published as *The Search for Order* in 1959. *Mind and Matter* was undertaken as an essay on the history of chemistry (in the widest sense) in response to my developing interests and the perception of a need by teachers of chemistry. By the time of its appearance, the intellectual climate in the United States was no longer the same. The perception of public education for a participatory democracy had been changed dramatically. Both books, after translations and reprintings here and abroad, gradually dropped out of print and have been unavailable for nearly the last fifteen years.

I do not think that the significance of the worldwide student disorder of the sixties has yet been properly analyzed. We tend to regard it parochially. In the United States we thought only of our own social and political causes–all the protest and dissension over the Vietnam War, the women's movement, Black liberation, and civil rights–labelling them, as we label all threat of change, as radical and alien. But the sixties were a period of worldwide student unrest, and universities erupted in Prague and Amsterdam and Paris and Sydney–in Peking as well as Berkeley and New York and Durham, New Hampshire. The buildings of the University at Hokkaido were completely wrapped *à la Christo* in chain mail when I saw them in 1970. The humanistic faculties of the University of Barcelona were sealed off by the dictator Franco's bayonets on the same day that I lectured at the mineralogical research laboratories. Creature of the state, the work of the laboratories was the work of the state and could not be interrupted. An anonymous graduate student of mathematics at Berkeley identified the true focus of the disorder. He was being used, as we all were being used; his mathematics was being used in the service of the Juggernaut. The sign over the entrance read "Plato's Academy," but the reality within was Sparta's army. Rationalism was perverted. Chemistry was employed for the invention of napalm, bacteriology for the creation of new diseases. Hiroshima had not been just a monstrous aberration but a foretaste of the future. In a terrible kind of ultimate irony, humanism was enlisted against humanity. It was for this that the students of Paris toppled De Gaulle and Prague turned to Dubček, and even here in New Hampshire, on the fringes of American society, the university teetered on the brink of violence. All over the United States, educational authorities, unwilling or unable to satisfy the clamor for social justice, yielded up the curriculum instead. Reason and science were branded as culprits. History was irrelevant. The Vietnam War ground on to its ignoble finish, but the old system of exams and requirements was irreversibly altered.

Last year, a quarter century after its first appearance, *The Search for Order* was republished (under the title *The Evolution of Physical Science*). Now *Mind and Matter* is to reappear. The crisis in education in science was never resolved. It has in fact deepened. The pattern of the subservience of science to the political-military-industrial complex that I saw in Franco's Spain is now the pattern of the American universities. To the insecurity of immense stockpiles of nuclear weapons, we have added the insecurity of the proliferation of nuclear power plants. Thirty years ago when members of the Society could fit at a single round table, it was a struggle to win acceptance for the history of science as an academic discipline, but it was a common problem and we met it successfully with a shared enthusiasm. Most of us went to great pains to make our work "user friendly." Today the Society's annual banquet requires a hall, and all too frequently the reputation of the professional historian of science is inversely proportional to the accessibility of his or her work.

This book represents my understanding of the progress of chemistry at the time of writing. Reviewing it is an opportunity to reflect upon the scientific developments of the two decades since. It seemed to me then that science was a search for order in the natural world and that it grew out of an ancient faith in a predetermined symmetry of pure thought and nature – the symmetry of mind and matter. Fascinated from the beginning by the complementarities of order and disorder, by the rebellion of the polytypes and snowflakes against the strictures of the international crystallographic tables, I believe that recent analyses of the fractal geometry of nature prefigure a fundamental departure in the received scientific perception of the natural world. Benoit Mandelbrot's mathematics describes an Anaxagorean world and present challenges to thermodynamics and the fundamentals of natural philosophy as profound as any of the present century. Unable to predict or foresee this in the slightest, I am pleased nevertheless that the reader of my twenty-year-old synthesis will find no impediment to the acceptance of these or any other inspired reorientations of the axioms of science. On the contrary, my conviction originally, as now, is that the study of the history of science should inculcate in all of us a consciousness of the frailty of our deepest beliefs and the humility to recognize our true position on the beaches of the immense ocean of truth.

Newfields, N.H., 1987

TO MY STUDENTS

MIND AND MATTER

1

From Speech and Fire to the Age of Iron and Letters

. . . AT NAUCRATIS, IN EGYPT, WAS ONE OF THE ANCIENT GODS OF THAT COUNTRY, THE ONE WHOSE SACRED BIRD IS CALLED THE IBIS, AND THE NAME OF THE GOD HIMSELF WAS THOTH. HE IT WAS WHO INVENTED NUMBERS AND ARITHMETIC AND GEOMETRY AND ASTRONOMY, ALSO DRAUGHTS AND DICE, AND MOST IMPORTANT OF ALL, LETTERS.

—SOCRATES*

By chemistry we understand the science of matter and the processes involved in its combinations and dissolutions. The distinctions between chemistry and the other sciences such as biology are somewhat arbitrary, especially if we consider that living creatures are physical and chemical systems and that vital processes involve the combination and dissolution of matter according to physical law. It is equally clear that processes which may be entirely outside of science, or at least of the physical sciences, play a determining role in the evolution of chemistry. I am referring to historical processes. The spread of a particular population of ants or grasses we choose to regard as entirely determined by rigid physico-chemical law, but the westward migration of modern religious dissenters—the Doukhobors or the Mormons—we regard as culturally and psychologically motivated, as indeterminate expressions of the

* Quoted by Sarton. (For complete references, see Bibliography, page 279.)

human free will. Our science of chemistry deals with materials. The history of our science is not so tangible or direct a study.

Between the earliest humans of our species carrying on within their bodies the complex biochemical and biophysical processes of the higher mammals exactly as we do today, and the highly articulate self-conscious citizen, carefully controlling his environment with electrochemical heat pumps, dressed in synthetic fibers, multiplying his strength by powerful engines, his speed by jets, regulating his diet by vitamins and controlling his caloric intake, even beginning to influence his genetic characteristics, lies the evolutionary gulf which is the history of mankind but is especially the history of science, including the history of technology and the history of chemistry. At the time of this writing (but not perhaps in a few years) we are still the same species as those fur-clad artist-hunters living on the fringes of the great glaciers who climbed down through the caves of Montignac to paint the record of their hunts and the strange species, the wild ox, the mammoth, the bison, with whom they shared their world. We are probably closer to them than they were to their (and our) remotest ancestors—the first of our species, the cave-dwelling food gatherers. Ben Franklin defined man as the tool-making animal. At the expense of pith, let us add to this, the *chemist*-animal.

Of course, man is not the only animal to utilize chemistry for his own purposes. Insects take pains to store and preserve food, sometimes fermenting it. The spider secretes a synthetic fiber with strength which Du Pont has not yet been able to equal. The squirrels in the New Hampshire woods intoxicate themselves on the apples fallen to the ground in neglected orchards, and they bite the buds of the maple trees in the spring to permit the sap to ferment. We must allow that this is chemistry and that there is a straight-line progression between the ant fermenting leaves underground for the fodder of the insects living symbiotically in his nest, and the larger scale ensilage activities of the dairy farmer or the highly sophisticated efforts of modern biochemists to convert sunlight and chemicals into food and oxygen aboard space ships and submarines.

Men of our genus, the *Hominidae,* have been on earth for

millions of years. The evolution of species which gradually led to the development of the hand, the eye, and the brain, with which we differentiate man biologically from the animals, led also to cultural differentiation—to the first use by men of suitably shaped stones and sticks for weapons or tools and, beginning perhaps 500,000 years ago, the first chipping of a stone or scraping of a stick to produce by craft what the environment had failed to supply. Surely with this hand, this eye, this mind, came the first language. Man cannot be defined in purely biological terms; as Franklin insisted, the essence of mankind is cultural. It lies in the things with which we are associated, and also the words we use to communicate. Descartes wrote that the distinction lay in the human soul and that the possession of language was the evidence of the spirit and the warranty of humanity.

Cave deposits and stream gravels which we know to be approximately 500,000 years old contain the partly modified stones which are the first evidences of cultural man, the large-brained, the far-sighted, the nimble-fingered, and the speech- and fire-maker. He is four interglacial periods before us. About four-fifths of the next 500,000 years were to pass before the appearance of our direct ancestors, the first of our species, *Homo sapiens.* Millions of years had been required for the transition from occasional user of improvised tools to maker of tools, user of fire and speech. Hundreds of thousands of years were to pass before the rude pointed flakes and sharpened animal bones left by Peking man were to develop into the spear- and arrowheads of *Homo sapiens.* Of the last hundred thousand or so years that our species has dominated the planet, almost nine-tenths were required to make the transition from the first *Homo sapiens,* gatherer of food and cave dweller, to the neolithic farmer, producer of food and builder, with whom human history proper began. As for civilization, the production of electricity for useful purposes is little over a hundred years old. Our conscious use of nuclear energy is twenty-four years old. This should perhaps lend us perspective.

In the Greek legend, Prometheus, admitted to Olympus by

Athena, the goddess of science, stole fire and became the first chemist. For having bestowed the fire of the gods on mankind and thus enabled man to raise himself to the level of the gods, he was condemned by Zeus to remain forever chained to a rock in the Caucasus with a vulture tearing at his liver. Today we can only guess in what manner the legend reflects the beginnings of man's control over materials. Prometheus was said to have obtained the first fire from the sun. The earliest manlike creatures must have obtained their fire from blazes accidentally started by lightning or by contact with glowing matter from active volcanoes. Peking man, it is known from the charred bones found in his caves, used fire. Paleolithic man of our species left behind iron pyrite and flint nodules with which he struck the sparks to light his fires.

With fire the entire range of technology becomes accessible. The whole of the plant and animal kingdoms, with a very few exceptions, became potential foods for man once he learned to soften them or chemically alter them by cookery. Over a period of thousands of years our ancestors graduated from hunting the wild cow and plundering her udders to penning her in and milking her. With fire, they learned to bake clay into containers, and with storage, they gradually mastered the biochemical processes of cheesemaking, of controlling the souring of their gathered milk and fruits. Over the millennia they learned to preserve their meat with salt and smoke, to soften furs and skins with alum and oak bark, to flavor their foods with plants like the onion, to fortify their beers with fermented honey.

The chemical technology of paleolithic men was not at all slight. We know that they used oil lamps consisting of a wick in a shallow depression carved in a stone to illuminate the caves in which they lived and painted. We know that they mined yellow ochre, hematitic clay, manganese, and lead oxides to be mixed with animal fats, water, urine, possibly even plant gums, for pigments and cosmetics. The beginnings of pottery date to the last Ice Age, when mammoth hunters of Moravia baked images of clay. Paleolithic man mined the soft chalks of England for fresh flint nodules with which to construct his tools. Here began the twisting of fibers and hairs into fish line and rope for snares or tethers, which culminates in modern textiles. To soften furs and skins for clothing required the rudiments of the tanning process, beginning with chewing to break down the fibers, in the manner of the Eskimo.

Perhaps not much more than ten thousand years ago our ancestors took to constructing artificial shelters, and spread from their caves to other environments. Man the hunter-gatherer became man of the New Stone Age, the settler, learning to grow grasses for his edible seeds, to domesticate animals, to soften the flax plant by retting and to weave the fibers into clothing, to fire pottery, to brew beer, to leaven his bread with domesticated strains of yeast, to tan furs and skins with alum and oak bark, to caulk his boats with pitch, to dye, to paint, to perfume. The world's population must have multiplied many times over. In secure and fertile river valleys, crude neolithic villages coalesced into cities. Urban society appeared, with the possibility of occupational specialization. At the same time the new science of metallurgy made its appearance, and with it, the Stone Age merged into the Bronze Age. Writing was discovered and history and civilization began.

Up to this point the history of chemistry and the general civil history of mankind coincide even in terminology. The stone ages, ancient, middle, and new, give way to the bronze and later the iron ages. But with the advent of civilization, our interests diverge from those of the civil historian. We shall refer to the period from the beginnings of civilization through the rise and decline of classical civilization in Greece and Rome as the Age of Letters because of the conviction that, just as the biological emergence of human vision, hand, and brain 500,000 years ago was necessarily linked to speech and to tool- and fire-making, so there is a causal connection between the invention of writing and the technological explosion that accompanied the birth of civilization.* By writing, in its broadest sense, we

* One significant exception is the ancient Inca civilization, which flourished without writing. But even the Incas employed the "quipu," a kind of code, chiefly numerical, in knotted cords. Other Pre-Columbian civilizations, particularly in Central America, had well-developed hieroglyphic systems.

mean the development of an artificial memory. With writing the experience and skill of one man may be made available to others remote in time and space. Before the invention of writing, each generation was obliged to relearn painfully all the lessons of the past. The diffusion of skills and knowledge was restricted to direct observation or speech. We may imagine the combinations of circumstances that would be required before a useful invention employed by one people—for example, the roasting of pebbles of brown-stream tin in a charcoal fire stimulated by a bellows for the production of bronze—could be adopted by another. The process would have to be observed in detail and understood, not by a passing hunter or trader or slave girl, but by a skilled artisan already familiar with aspects of the process. And even more than its parent, speech, writing is a process of abstraction. The map takes the place of the terrain. The account book takes the place of the warehouse. The story of Joseph and the sabbatical years of plenty and famine illustrates the early use of literate abstraction for sustaining a large population. With writing, abstract thought makes its appearance. The foundations of the system of cumulative knowledge, which we call science, are possible.

Of no less importance than the process of writing itself were the related inventions of inks, of papyrus made from a river sedge, and of the glues that formed papyrus into volume rolls. Religion, medicine, and science in ancient Egypt were preoccupied with the problem of decay, which gave momentum to their scientific endeavors. Disease and death were transitional stages to a universal corruption, to be resisted by the compulsion of magic and arts, and by religious submission and appeal.

Typically, Egyptians forced immortality on their rulers by preserving their corpses so skillfully that many examples of mummies survive in our museums today. They employed light (naphtha) and heavy (asphalt) petroleum products, ordinary salt, natron (carbonate of sodium), niter, sal ammoniac, potash, alum, aromatic resins and gums, wines and vinegars. The elaborate mummification process involved disembowelment with cleaning and preservation of entrails with wines, vinegars, and salts, oiling, desiccation, sanding, swathing in bitumen-soaked cloth—rituals of pickling and salting carried over from the preservation of fish and meat. From the firing of clay for ceramics came the discovery of fusing surface glazes and then of glass itself, with an elaborate mineralogy for the coloring of glass. Ancient society understood baking, brewing, and wine-making but not distillation. Ointments and unguents, balms and cosmetics were prepared by using fats and oils to absorb perfumes from flower material, sometimes with cooking, steeping and so on. Gum and oil resins were burned as incense.

Paleolithic man's use of metals had been largely confined to gold and silver and occasionally copper, to those metals which were found in native (metallic) form and therefore would attract attention by their unusual luster. These metals were used as jewelry but also as small implements such as pins and needles. In the Incas' Peru, trepanning instruments of gold and silver, significantly resistant to corrosion, were successfully employed to repair skulls fractured by stone battle axes. The ores of copper would have early attracted attention by their vivid colors and could be smelted at relatively low temperatures with charcoal and a saline flux to reduce them. Castings of copper are known from the fourth millennium before the modern era. To be useful, copper implements had to be hardened by suitable combinations of hammering (cold and hot working), annealing (heating and slow cooling), and quenching (quick cooling). Although rarely used, the mineralogy of ore discovery, the refining and smelting processes, casting and tempering, were known to prehistoric man. But it was the introduction of bronze (alloys of copper chiefly with tin) which marked the beginnings of true metallurgy. Bronzes are prepared at lower temperatures and are more easily cast and worked than pure copper.* Bronze will hold an edge and is therefore more suitable for tools and weapons than copper. It is harder and more durable and will serve for axles and bearings. Because other minerals are added to the ores of copper, bronze-making increases the production of metal. The Bronze Age

* The addition of about 50 percent tin lowers the melting point from 1083° C. for pure copper to about 500° C. and raises the tensile strength about 20 percent.

coincides with the beginnings of civilization and the Age of Letters. Lead, in the form of statuettes or jewelry, was also known in the earliest of the river valley civilizations.

Iron, the fourth most abundant element in the earth's crust (after oxygen, silicon, and aluminum), is by far the most useful metal known to man. Although it was occasionally found in native metallic form (chiefly in meteorites) and was known in this form to paleolithic man, it appears on the human scene in quantity not much more than three thousand years ago. The ores of iron are hematite (red oxide), limonite (rust and bog iron), siderite (the carbonate), and magnetite (black oxide). Although widespread, they are not conspicuous or obviously metallic in their properties. These ores, when mixed with charcoal and flux and roasted in a hearth, yield a "bloom," a stony, porous, cinderlike mass which must be heated and hammered until the minute fragments of metallic iron dispersed through the stony slag are fused together in a small lump of wrought iron. Before this wrought iron can be steeled, or hardened, it must be worked by hammering. It must be carburized (heated in contact with charcoal), quenched, and possibly annealed. If any of these processes are omitted or improperly performed, the iron becomes inferior to bronze for neolithic purposes. The casting of iron (reducing it to a liquid state) is an even more difficult affair, yielding a brittle, unworkable product. Until the discovery of carburization by the Chalybites of the Armenian mountains in the second millennium before the modern era, iron was an impractical material. But with this discovery and its spread, metals became universally available to mankind. R. J. Forbes calls iron the democratic metal.

The *Iliad* is an account of an episode in the decline of the older Bronze Age civilizations before the onslaught of the iron-bearing peoples out of the Northeast. A thousand years later Caesar's legions swinging their short iron swords in good order were to subdue the Early Iron Age Britons in one short summer, just as these bronze-bearing peoples had previously driven the earlier Bronze Age inhabitants into the outer islands and the remote mountains. With iron came the discipline, the organization, the *civilization* of Rome. Behind the short swords stood the cleared forests, the drained swamps, the fittings for chariot and galley, all the technology of the New Iron Age. In the first millennium of carburized iron, the basic framework of our history (Greco-Roman), our theology (Hebraic), and our reasoning (Aristotelian) were fixed.

Even as the use of writing accompanied the grouping of peoples into city-states and the technological revolution that made empire possible, so this process was accompanied by a growing self-consciousness, an introspection that compelled the Hebrews on the one hand to search for final purposes and the Greeks on the other to seek out a mechanism. It was not yet possible to separate science—an understanding of the workings of the natural world—from theology—a suitable accommodation to the supernatural world—or from philosophy. The abstract monotheistic vision of the Hebrews was also intergrown with Eastern concepts of natural history. The mathematical and atomistic concepts of the Greeks arose out of and for the service of supernatural considerations of meaning and purpose in the world.

By the beginning of the Christian era, brasses (zinc and copper) had been used for a millennium in Judea. Arsenic ores and metallic antimony were known. Lead was used for plumbing and shipfitting. Alloying (the preparation of bronze and brass with specific properties by suitable mixtures of molten metals), casting, amalgamating (mixing with mercury), assaying, and even plating to counterfeit gold, were understood. In the well-known story of Archimedes, that philosopher is supposed to have observed the rise in the level of the water in his bath as he lowered himself. In this way he discovered the property of gravity in species or relative weight, which he used to detect the admixture of base metal in the golden crown of the tyrant Hiero. The touchstone, on which the purity of gold was determined by comparison of its color with known samples, was another means of analysis.

The natural selection which in other living things developed specific organs or limbs for specific purposes—horns and teeth for combat, hoofs for running, spinnerets for web-making—led, in mankind, to the eye set in front for depth perception, to the hand and the brain. As these evolved together, a cultural evolution accompanied them which culminated perhaps a hundred thousand years ago in *Homo sapiens*. In place of teeth and claws, the brain had the power of abstraction, language, and memory with which to direct the hand and the eye to fashion tools and to alter matter. The appearance of man, the maker of tools, fire, and words, was the last geological event. His gathering in large centralized groupings—cities—about five thousand years ago was made possible by a rapid evolution of culture which culminated in the development of writing, a first-stage artificial brain.

At the same time an enormous sophistication in the techniques for altering matter (chemistry) had appeared. The range of sister forms of life which man was able to turn into digestible food had been widened immeasurably. Cooking, tanning, brewing, dyeing, preserving, salting, smoking, pickling, bleaching—all made the new civilization possible. At the same time, and just as man had developed from a hunter-gatherer of food living in natural caves into a producer of food living in shelters of his own design and construction—with the beginnings of civilization—from a gatherer of stone, wood, furs, and fiber, he became ceramicist and metallurgist, converting earths into utensils and tools which in turn made possible the further expansion of his civilization.

Nor was he content at this or at any other stage of his history with merely improving his ability to survive. The paintings in the Lascaux caves may have had magical purposes of insuring success in the hunt or promoting fertility among the food-beasts, or they may have served as tribute to supernatural forces, but there is a sensitivity and a delicacy in their execution which marks great works of art. Five deer swim perpetually across the upper wall of the Lascaux nave, the bold black strokes in which they are outlined emphasizing the emotion of their flight. Man celebrates his prowess in the hunt. He creates anew the animals of the upper world, fascinated by their fertility. From the beginning his tools are decorated. The earliest ceramics show an attention to color and form. Both men and women adorn themselves. The history of mankind is not simply the record of man's progressive dominance over the world but also of man's joy in life and of the complex of passions and psychological and intellectual drives which distinguish him from the animals.

With the discovery of the carburization of iron, archeology yields to history. Peoples become nations. Among the shadows we begin to distinguish individuals and to note the shadings of their personalities. Prometheus is insubstantial, like Moses a symbol, a demigod, hardly a man. Pythagoras, or better, Zeno, leaves a clearer picture. The eye, the hand, and the brain have not changed but the culture has evolved. It is no longer dependent on natural selection, on the accidental survival on a statistical basis of the few individuals gifted with superior characteristics. In one summer of blood the Britons bowed to the Roman short swords and the bloomeries which made them. With the Iron Age came philosophy (and within this heading are included theology and science). Society at the dawn of the modern era was already literate, self-consciously cultivated, and by its lights, highly moral. If Caesar could put out the eyes of Vercingetorix and order him tortured to death, he could also record with admiration that the same Vercingetorix surrendered to save his people. The empires that spread soldiery and slavers into the forests also spread civilization and law. Ethics and morality were as essential to civilization as the grain ships bearing tribute from the frontier settlements.

2

The Idea of Matter

IN PHYSICS, FROM THE ANCIENT PHILOSOPHERS ONWARDS, A TENDENCY TOWARDS THE DISCONTINUOUS HAS CONSISTENTLY MANIFESTED ITSELF IN THE SHAPE OF ATOMIC AND OF CORPUSCULAR THEORIES. . . .

—PRINCE LOUIS DE BROGLIE, 1937

Early in the sixth century B.C. the rebel Zeno was tortured to death by the tyrant of the Greek city-state of Eleas, in southern Italy. A follower of Parmenides the Eleatic, Zeno is remembered for the famous paradoxes he proposed, not so much for the confusion of his countrymen, but as a serious argument in the great philosophical debate which the Greeks had been conducting for centuries. Zeno argued for Parmenides that the world was static, timeless, and change purely an illusion. Even motion was illusory. An arrow could not move in flight because time consisted of a series of instants or points with neither beginning nor end. In one instant of time, an arrow could not move, and therefore in all time the arrow could not move. Similarly, an arrow could never reach its target. In order to reach the target it would have to cover half the distance to the target first. This would leave another half-distance to be covered. No matter how many times it covered half the distance to its target, there would always be half a distance remaining. In the first paradox Zeno assumed that time was made of discrete, irreducible points. In the second paradox he assumed that distance was continuous and could always be subdivided. The first assumption was *atomistic,* the second *continuous.* This method of logical dialogue, of tracing the implications of conflicting assumptions, was the earliest use of dialectics. Zeno argued that Achilles, the swiftest runner of Greek mythology, would never catch a tortoise. For while Achilles covered the distance to the spot from which the tortoise started, that reptile, however slowly, would move some distance away. Achilles would still have a small distance to go, and in the finite time that it took him to cover that distance the tortoise could move again. Since Achilles would always require time to cover the small distance remaining to the tortoise, the tortoise would always be able to stay ahead.

Zeno's paradoxes were not meant as scientific observations about the world, or as a naturalist's theory of tortoise motion, but as a serious objection to those trends of Greek philosophy which thought to explain the world either in terms of number and form or in a more material sense. Zeno was trying to say that there was but one *Being,* timeless, changeless, a superatom or monad of existence—divine without being godlike; in fact, Being without change—necessarily without life. Such was the ultimate nature of reality. The Greeks were exercised over the problem of cosmology, the sum of all things. They asked only the largest questions, they found only large answers. "What is the nature of existence?"

For their contemporaries the Hebrews, the question was the purpose of existence. "Where do we come from; what are we doing here; where are we going?" Theirs was a spiritual and political cosmology. Job cries to his God for justice, and men reading the Book of Job tried to account for the presence of evil in the world. To what end were the sun, the moon, and the stars created? For what purpose? Is death then final? What is the relationship of man to God?

The English philosopher Collingwood says that there is a curious turn to the Greek mind, which alters these questions to others which have a different ring. The Ionians, according to Aristotle, when faced with the question "What is nature?" at once changed it to "What are things made of?" For them the problem of nature and being became the problem of "finding the primary unchanging substance which underlies all the changes of the natural world." This *monism,* or belief that a single reality underlies all existence, was essentially secular for the Ionians. It is the origin of the strongly naturalistic influence in Greek thought.

The Greek myths begin with a great disorder, a chaos which is both male and female and out of which emerge the elder gods whose purpose is to begin the process of ordering and rationalizing which we term history. The elder gods—monsters, giants—are in turn destroyed by their own children, defeated by the very progress in ordering existence which they themselves began. So cosmology is conceived as a progression from turbulence to order, from chaos to reason, from primal atheism to a universe of logic. In the beginning is the stuff of existence and out of it come the gods.

For the Hebrews existence begins with God; light is created out of darkness. The darkness needs no creation since it is simply the absence of light. "And the earth was without form and void; and darkness was upon the face of the deep. And the Spirit of God moved upon the face of the waters." Like the darkness, the ocean deep needed no creation. Light was created by the will of the Spirit, and earth by the gathering together of the waters. The Firmament was created as a protective cover; above and below it are the universal waters.

The Ionian Greeks of the sixth and seventh centuries B.C. saw in these universal waters the underlying element, the substance from which all things must come. According to Thales of Miletus, life itself begins with a seminal fluid and is nourished by fluid. The cosmos is a living thing, the universe is a gigantic animal. Not for twenty centuries would it again occur to anyone to divide the world of nature so sharply into the living and the dead, the organic and inorganic kingdoms. "All plants," Johann Kepler wrote in 1609, "are merely branches of that general drive which reigns in the earth . . . as the formative drive of water is to fish, as that of the human body is to lice. . . ."

For Thales' pupil Anaximander, the underlying element was completely without properties, a cancellation into nothingness. As wetness annihilates dryness, or in modern terms, as a positive charge encounters a negative, Anaximander thought of the world as made of opposites, hot and cold, light and dark, arising by differentiation of the underlying stuff. In part it is motion operating on the formless, limitless stuff of existence which brings about the formed, measured, ordered world of properties. The underlying substance was divine, in the sense of being above and beyond nature. The Anaximandrian concept is not unknown to modern physics, in which subatomic particles such as the electron and positron collide with the annihilation of their charge and mass.

For Anaximenes, pupil to Anaximander, the underlying matter was vapor or air; all other matter arose from the air by a kind of condensation, as water comes from the breath by condensation on a cold plate.

Now this is essentially chemistry, since it seeks to explain natural phenomena by a kind of analysis. Milesian monism seeks a single substance from which all the diversity of material experience can be derived. As it endeavors to account for this diversity by reasonable explanation, it verges on science. An explanation for the diversity is sought in motion by Anaximander, in condensation and rarefaction by Anaximenes. If they are not experimenters they are at least observers, recording for their speculations the transformation of matter—the earth into plants into animals, the fuel into vapors and fiery motion, living breath into water into crystalline ice. They are materialists in that they refer the world of nature back to a universal substance, secularists in that the substance is sufficient in itself, yet pantheists in that just as it has never occurred to them to separate matter from spirit, so it has never occurred to them to doubt that the cosmos lives, permeated by mind. When we speak of "nature" we share with the Greeks this belief in a living, thinking world. It is this mind, this nature, which is the source of that order by which we are surrounded, and the comprehension of which is both philosophy and its subdivision, science.

Pythagoras of Croton was not so much concerned with the underlying substance as he was with the logic of the differentiation. Like the Ionians before him he thought of a primary fundamental substance having the attributes of both matter and space. "The first principle of all things is the One . . . which is cause." To account for the differentiation into materials as we know them, Pythagoras imposed form (by which he meant number and geometric form) on his "matter for the One." From his numbers came geometric figures; and from these, bodies apparent to our senses. "The four elements of fire, earth, air and water are the elements of sensible bodies transforming one into the other, and out of them comes to be a *cosmos;* animate, intelligent, spherical, embracing the central earth, which is itself spherical. . . ."* Just as a circle, a triangle, and a square differ not in what they are made *of,* but purely in their form, so the Pythagoreans held that different materials differed only in their form, that the essence of material things was in their geometric form.

Figure 1. Johann Kepler's Table of the Elements
The five Platonic solids are equated with the five elements; from *The Harmonies of the World,* 1619.

* Cornford, 1957b, p. 3.

We argue with Pythagoras: the steam from a boiling vessel and the lump of ice lying inert on the ground are made of one and the same substance: not water, which would be something else again, but hydrogen-oxygen. How is it, then, that the properties of the two are so vastly different? Pythagoras would have answered that the difference is in the geometric form of each. The modern student of ice and steam would give an answer essentially Pythagorean: the difference lies in the arrangement and distribution of the primary matter. In ice the molecules are close together in an orderly arrangement, like bricks in a wall. In steam the molecules are far apart in violent random motion.

Pythagoras regarded the world as a harmony, by which he understood a purely numerical differentiation. Harmonies in music occur when the lengths of strings or pipes bear certain simple numerical relationships—1:2, 2:3, 3:4, and so on. The essence of the qualitative differences of things is *mathematical.* (When we see why this is so significant we shall see that the modern scientist is far closer to Pythagoras than we may realize.) For Pythagoras the essence of diversity in nature is mathematical form. The essence of things as they are is therefore supremely *intelligible;* not magical, not mysterious—divine, perhaps, but in the sense that form is ideal and therefore divine.

The attempt to grasp the nature of existence by the effort of pure reason reaches its highest point in Plato of Athens (c. 430–347 B.C.), the founder of the Academy and the master of Aristotle. In the thought of Plato form and number become the purest examples of *ideas,* transcending the material world, which loses its aspect of reality. To illustrate the unreality of that material world which manifests itself to our gross senses, Plato wrote the parable of the cave. He imagines prisoners in a cave chained so that they can see only the shadows and images cast by firelight on the wall. One of the prisoners is released and confronted with the world of trees and men, the heavens full of stars, the moon and the sun. Returning to the cave, blinded again by the darkness, he is unable to persuade his fellow prisoners of the reality incomparably more glorious than the imperfect shadows which fill their narrow world. "Just like ourselves," Plato says; the experiences of our senses are but the pale reflection of a greater, glorious truth. Behind the sense impression lies the pure idea. The chalk circle on slate, the wheel, the square table are the rough embodiments of the ideal form, the circle, the square. The idea is real. The direct object of experience is illusion. In his youth Plato is said to have studied the doctrines of Heraclitus with the philosopher Cratylus. Heraclitus, last of the monists, saw change itself as the one constant in nature. Fire, since it is always changing, was therefore the primary element. But if everything was always changing, it seemed to his follower Cratylus in his despair, no statement could possibly be true, and therefore he abandoned speech and restricted himself to wagging his finger. In this Plato did not follow Cratylus, for although Plato saw only illusion and deception in sensory experience, he nevertheless maintained it is all we have. Nature may be always in flux, but this is no reason to abandon the attempt to distill from nature that which is intelligible—the form.

The influence of this current in Greek thought is most direct in astronomy. Encouraged by Greek ideas of the divinity of the stars and planets, Plato's pupils and their intellectual descendants down to and including Kepler and Galileo searched out the geometric regularities of the heavens.

The apparent chaos of the skies, the disorderly and retrograde motions of the planets, led Plato to propose his famous problem: to reconcile the apparent motions of the planets to a combination of uniform motions in perfect circles and thus to "save the appearances." This view of the aim of natural science is very close to the mathematical formalism of modern theoretical physics, which may see its task as the formulation of mathematical expressions conforming to certain preconceived restrictions and ultimately connecting with observations. But is this end result really so unexpected? Do we not think as we do and fashion the kind of science we do precisely because our civilization has been set in this path from the beginning by Pythagoras and Plato?

Empedocles of Agrigentum (fifth century B.C.) was supposedly the formulator of the doctrine of the four elements—earth, water, air, and fire. He conceived of them as indestructible, accounting for the properties of matter by their mixing and for the transformations of matter by their separation. Despite the attempts of the monists to base the material world on a single elemental substance, most of the Greeks, including Plato, accepted the idea of the four elements. Earth, water, air or spirit, and fire or light are the common fundamentals of all early cosmologies. We know that Genesis* presupposes the Spirit of God and the existence of a formless void, earth, brought into the newly created light by a gathering together of universal waters. Water, earth, and Spirit are the implicit elements of Creation, with a fourth, light, and a fifth, firmament, created in the beginning. The earliest Greek formulations also assumed three elements, presumably by lumping together fire and air. In India the doctrine of the elements is at least as old as Buddhism (sixth century B.C.). At first these were (probably) fire, water, and earth, with air added later and the fifth element, the ether, still later.

> Oh Earth, my mother, Air, my father, Oh Fire, my friend, Water, my kinsman, Space, my brother, here do I bow before you with folded hands!†

In ancient China the first formulation of a doctrine of the elements was made by Tsou Yen some time between 350 and 270 B.C. Water, fire, wood, metal, and earth were thought of as elemental properties or processes. They were associated with the planets,‡ with the seasons, and with the directions of the compass, as part of a numerological mystique that attached special significance to the number five.

From the limited evidence it is impossible to reconstruct the connections between the Chinese, Indian, and Western doc-

---

* Part of a fourth-century B.C. compilation from sources going back to the ninth century B.C.

† Quoted in Basham.

‡ The Chinese associated fire with Mars. In both Ptolemy and Pliny the quality hot or fiery is associated with Mars, the quality cold and dry with the element earth and with the planet Saturn. The Chinese associations are Saturn-earth, Jupiter-wood, Venus-metal, and Mercury-water.

trines for the elements. Yet some authors insist on independent origins for the Indian and Greek concepts, which are the same.

> Any dependence of the Greek upon the Hindu or of the Hindu upon the Greek is not on that account to be thought of, for the reason that the order of sequence is different in that with the Greeks air comes between ether and fire.*

Earth, water, and spirit or air or fire are the three elements common to all the ancient cultures of Europe and Asia. The historical record is too limited for us to infer either a connection or a theory of independent origin, but it is at least reasonable to attribute the concept of elements and the formulation in terms of earth, water, and spirit to preliterate cultures. The differences between the Chinese and the Indo-European five-element doctrines may be evidences of the later independent evolutions of these ideas in China and Indo-Europe, and of the unified cultural evolutions of India and Greece into classical times.

It would appear a rash venture indeed to attempt any history at all of preliterate and therefore prehistoric man that presumes to touch upon his ideas, but writing is not the only means of communication in time. The tools, the weapons, the burial ceremonies, give us insight into the beliefs of prehistoric man independent of any extrapolation from the practices of modern primitive peoples. The Stone Age Patagonians who so impressed the youthful Charles Darwin had a history neither longer nor shorter than his own. Neanderthal men were buried with tools and weapons and small supplies of food. Sometimes they were buried in the caves in which they had lived. Was this done in the belief that their presence would protect the living? Sometimes huge stones were put over them, perhaps to protect their families from *them;* sometimes their hands and feet were bound to prevent their roaming. The picture of man's early beliefs in the supernatural which we reconstruct from these evidences is necessarily general, but it enables us to say with assurance that men of the genus *Hominidae,* even of species earlier than our own, believed in life after death, feared

* Quoted in Basham.

ghosts, and in fact shared with us the human trait of an intellectual life. The cave paintings in France and Spain were the expressions of peoples whom we can classify only on the basis of their tools, and served purposes which we can only surmise. Why did they paint only animals, with rarely if ever a human being? Why is there no vegetation, no scenery, no landscape? If there were rituals connected with the paintings, why are the caves almost devoid of traces of use? What are the stick-like symbols which have been called traps or corrals by some? How did so natural and expressive a mode of art persist through more than twenty thousand years, through the cultural changes we call Aurignacian-Périgordian and Solutreo-Magdalenian, only to disappear and be supplanted by the far more primitive artistic techniques of the first civilizations?

Although we can answer these questions only with other questions, we can nevertheless draw a picture of a society deeply involved in activities which our society labels intellectual and which in our arrogance we assume to be restricted to our own times. Our intellectual leaders like children are aware of only the immediate past, and are therefore capable of telling us that perhaps 90 percent of all the scientists who ever lived are living today, or that the volume of printed material roughly doubles in each decade, or that our science doubles in every decade.

But a reasonable case might very well be made for the scarcely more outrageous assumption that the greatest part of this prodigious volume of effort is sheer noise, contributing little or no information of value. Would it be going too far to assert that all the truly important discoveries of mankind were the work of anonymous paleolithic geniuses, with their major development complete by the end of neolithic times?

The only animal to have been domesticated in historic time is the ostrich. Number, art, fire, clothing, music, were all inventions of paleolithic man. The domestication of the grasses, bread, wine, writing, the house, the boat, the lever and the wheel, the city itself—all were prehistoric discoveries, along with astronomy, arithmetic, and God. Are we really to equate the development of the seedless watermelon by our agricultural research laboratories with the domestication of corn by the neolithic American Indian? Is the cordless electric toothbrush an invention comparable with the sail canoe?

The cave paintings of Western Europe are still, at the time of this writing, a mystery. Did they serve purposes of magic, of religion, or perhaps some activity that played as significant a role in the life of that society as religion does in ours—an activity as rich and varied and certainly, with its span of tens of thousands of years, far more culturally stable? Certainly they tell us that the intellectual life of paleolithic man was no simpler than our own and that we stand today in the same ignorance of our racial past as those pioneers of seventeenth-century thought, Bishop Wilkins and Father Kircher, who first reasoned that the hieroglyphics of Egypt were an ancient writing and thus glimpsed the possibilities of history two centuries before the writing could be deciphered.

In this perspective, it appears reasonable to attribute to paleolithic man, before his separation into eastern and western neolithic societies, the concept of three interconvertible elements: earth, water, and fire or air or spirit. Not until the clearly defined age of the prophets—the time of Buddha and monotheism in India and the Near East, of philosophy and manners in Greece and China—is the doctrine of the elements written down and thus entered into history, with the differences so clearly linking Greece and India on the one hand, so clearly separating them from China on the other. In Greece and India the increase in the number of elements is accompanied by the idea of atomism with the materialism which it entails, while in China, in so many other ways more materialist than the West, atomism is never developed.

In the Platonic cosmology each of the elements is identified with one of the five perfect solids: fire with the tetrahedron; air with the octahedron; water with the icosahedron; earth with the cube; and the ether, the fifth element making up the heavens, with the pentagonal dodecahedron. These five perfect or Platonic solids are distinguished by having all sides, edges, and angles equal. In the Pythagorean mystique of number and form, the Platonic solids, as well as the numbers 7, 10, and 12, and the circle and the pentagon, take on a significance which our society reserves for fetishes. Lingering evidence of this respect for numbers may be seen in the skyscrapers built without a thirteenth floor and the flood of astrological literature in our daily press.

The sides of the tetrahedron, octahedron, and icosahedron are equilateral triangles. Plato proposed that the transformations between fire, air, and water could be explained by reconstructing any of these solids from the triangular sides of another. Similarly, transformations with earth (the cube) would come about by breaking down and rearranging the edges of the polygons. Clearly these are not elements in the sense of the modern chemical element. When we say that water is composed of the elements hydrogen and oxygen, we believe that we can decompose water into these two kinds of matter and nothing more fundamental, and that we can combine them again into water with no residue. We believe that nothing can be gained or lost in this experiment and we exclude from the category of thing, or matter, all products of our experiment about which similar and equally precise statements cannot be made (such as light and heat). The Greek idea is much closer to our modern concept of states of matter—solid, liquid, gas. A metal might be composed of a watery part (seen as it melts) and an earthy part, the calx or ash of a metal which has been heated in air. The steam from boiling water was considered an example of the transmutation of the elements. It is only when seen in this light that the efforts of the later alchemists to transmute baser metals into gold may be understood.

For Plato the basis to which matter can be referred is a kind of primary formless stuff—a surd receptacle, irrational and somehow evil. In the Pythagorean mystique, if number was to be identified with nature then the existence of the irrational numbers—numbers such as the $\sqrt{2}$ or π, which cannot be expressed exactly by any ratio of integers—was somehow a reminder of that original chaos from which the elder gods had emerged bringing order and light.

We distinguish two trends in classical thought, two trends which are quite contradictory. Both are based on observation and experience to some extent, yet neither is so based in the sense in which people frequently refer to the scientific method. In the popular sense, scientific method is the collection of facts (whatever is meant by this), their assembly into an orderly pattern, and then, somehow, the distillation from these of brilliant generalizations which become scientific theories or, with the discovery of more confirmatory facts, laws of nature.

The Greeks about whom we have been speaking up to now proposed an orderly explanation of one kind of experience, the experience by which we see one thing transformed into another —grain into meat, the child into the man, the wood into ash. They conceived of philosophy as a search for unifying, simplifying ideas, therefore they thought to refer natural changes back to one primordial matter. Yet they were essentially hostile to materialism. They placed matter and substance in some kind of secondary category, emphasizing that it is mind, or rather idea in the guise of form, which gives significance to matter. But there was another view which sought to explain existence and cosmology in terms of matter. For this more materialist view, matter or substance was the single underlying fact of existence, complete in itself, requiring no intervention of idea or spirit or god.

The two trends in Greek thought can be loosely termed Platonic and Democritean. The Platonic is essentially idealist. It is *continuist.* Aristotle (to whom we owe much of our knowledge of Greek philosophy and who may, for our purposes, be classified with the Platonic wing) says that nature abhors a vacuum. He cannot conceive of a space not everywhere permeated by some substance, however thin. Before we rush to scoff at this, it would be well to recall that by a vacuum Aristotle understands a space with *nothing* in it, whereas our laboratory vacuums are simply chambers from which the air has been removed but which may pass such things as light, electromagnetic waves, and gravitational force, and which may have properties such as resistivity. The world for this branch of thought is a plenum, full everywhere, without holes. The world is time as well as space—a continuous river of time down which floats a spatial plenum.

The group that we may consider together under the term Democriteans are the first true materialists. Democritus himself, a Greek of the fifth to fourth centuries B.C., considered that the only real things were atoms and the void or nothingness between them. The void makes motion possible and motion is itself the principal property of atoms. The primordial atoms are themselves continuous, full, and indivisible, and in themselves immutable. The diverse properties of matter and all phenomena are to be accounted for by the arrangements or positions which the atoms may take. In place of the single, all-encompassing monad of Parmenides, Democritus has an infinite number of minute monads. He believed that the individual atoms could themselves have form—little cubes or little rods or little crosses. The atomists are the opposite of *continuists.* They accept the idea of the void. They believe that substance is discrete, in small particles or pieces separated by void. Aristotle says that Leucippus of Miletus was an atomist in order to have the idea of void, which he insisted was necessary for motion to occur. Like Aristotle himself, Leucippus did not allow the arguments of Zeno to persuade him of the nonexistence of motion, saying that motion existed and therefore void must exist.

While neither of the two schools of thought was concerned with experience in the sense that we would today consider linked to experiment, nevertheless the atomists made a certain appeal to experience. If we pour wine into water, dissolve salt, or draw a deep breath, how can we explain these mixings except by the analogy of pouring water on sand?

Anaxagoras of Clazomenae (c. 488–428 B.C.) was not strictly an atomist, since he believed that substance (and probably also space and time) was continuous. He argued that any substance could be divided indefinitely into smaller and smaller pieces or eternal and incorruptible seeds, always retaining the properties of the substance. These small pieces, themselves indefinitely divisible but always retaining the same properties, Aristotle refers to as homoiomeres ("like parts"). The reader who takes a piece of common mica and splits it into sheets themselves splittable into ever smaller flakes is illustrating Anaxagoras' concept of homoiomeres. We shall meet something like them again in the history of chemistry with the Abbé Haüy in the nineteenth century. Many kinds of these seeds of Anaxagoras were present in substances, transformation being possible by the separation of these seeds.

This view is not the strictly atomic view, which would hold that there must be ultimate atoms themselves indestructible and indivisible or else the world would erode into an always finer dust and pass away. The purest atomism would ascribe to the atoms neither size nor weight nor properties, making them only ultimate material points. Democritus had held that color, taste, and smell were not properties of the atoms but secondary or subjective properties excited in the mind of the observer. He had ascribed to the atoms size and shape and perhaps even weight, at the same time considering that most of the properties of materials arise out of the patterns taken up as the atoms come together—an aggregation which he thought the atoms must accomplish by mechanically catching one in the other like so many hooks and eyes. Leucippus, an extremist, held that nothing exists but atoms and the void. Thought and emotion, as well as light, brick, wind, gold, water, and even heat and cold, were either atoms or derived from the motions and patterns of atoms.

We can now distinguish several possible kinds of atomism. First is the ultra-atomism in which we have only material points in a void, with properties arising only out of the aggregation and motion of the atoms. Then there is the idea of homoiomeric atoms which themselves have properties, such as size or weight or shape. There must be a very large number of kinds of these, since the world is so complex. To contrast the shapeless or spherical atom with the Anaxagorean homoiomeres we may coin the term "aniomere" (lacking parts) for it.*

Greeks adhering to either the Platonic or the Democritean ideas subscribed to the doctrine of the four elements, fire, earth, air, and water, which persisted into the seventeenth century in western Europe. They also identified the elements with

* "Aniomeric atom" is of course redundant, but in the same sense that the modern use of the word "atom" is a contradiction. "Atom" literally means indivisible, or without parts.

the humors of the body (blood, phlegm, bile, and black bile) and in turn with the five visible planets. To this day we speak of a *saturnine* or a *jovial* disposition for those whose humor is *bilious* or *sanguine*. In this way these ideas were linked with medicine, with astronomy, and with astrology. To the extent that these studies, and particularly astrology, rivaled religion, they were subject to continuous denunciation throughout the ages, a denunciation which leaves the scientist, even in modern times, the object of suspicion and mistrust.

In summary, the Greeks posed the problem of chemistry to account for the material diversity of the world in orderly and intellectually satisfactory terms. Inheriting the rudimentary concept of three or four elements, the monists attempted to reduce these to a single fundamental principle. The Pythagoreans identified them with form and number. In place of the one being—the superatom of static existence of Parmenides—the atomists explained material diversity with an infinity of atoms, the number, configuration, and motions of these giving rise to all phenomena. The Greeks accurately predicted that future ideas would fall into continuist or atomistic schools, and they provided logically consistent solutions in both categories. By demanding reasonable and above all mathematical explanations, they set the problems of future science. What the Greeks did not do—and what was not to be done until the Renaissance—was to understand the necessary connection between thought and practice, between philosophy and utility: the interdependence between the life of the mind and the requirements of living. With the Renaissance achievements and the Baconian emphasis on practice, science was to embark on an orgy of successful accomplishment. Today, in the maturity of a great technological era, science becomes introspective and seeks to find for itself more meaningful models than the simple mechanism of the Renaissance. As we turn deeper into the examination of our science, we find ourselves, with Plato and the Pythagoreans, apprehending the world as number and form, born out of some chaos which only the exercise of reason keeps temporarily at bay.

3

The Middle Passage: Classical Alchemy to Medieval Crafts

THE BODY . . . IS ONLY THE PLACE OF REFUGE AND SOJOURN OF THE SPIRIT . . . THINGS THE MOST STABLE ARE THOSE WHICH CONTAIN MOST OF BODY AND LEAST OF SPIRIT; SUCH ARE GOLD, SILVER AND ANALOGOUS SUBSTANCES. THINGS THE MOST FUGACIOUS AMONG BODIES ARE THOSE WHICH CONTAIN THE MOST SPIRIT; SUCH ARE MERCURY, SULPHUR AND ARSENIKON.

—DJABER*

The period from the time of Pericles (c. 450 B.C.) to the age of Constantine (c. 300 A.D.) saw the establishment of standards of culture which became canonical for all subsequent societies in the West. These were the patterns of thought and art and architecture, the attitudes toward government and law and particularly toward nature, which all subsequent societies have termed classical. The earlier Greek society was the Hellenic; the later, Roman-dominated culture is called Hellenistic.

At its height Greco-Roman civilization reached levels of accomplishment so imposing as to disturb our complacency as moderns. Yet the simplicity and much of the charm of earlier stages of civilization were lost in the great commercial centers of Imperial Rome. In philosophy the clarity of the Platonic dialogue of Periclean Greece gradually yielded to the mystical

* Quoted in Stillman.

neo-Platonism of the Egyptian Plotinus (c. 204–270). Alchemy made its appearance. Its principal objective, originally the counterfeiting of gold, became the transmutation of base metals into precious metals. What had been, in the earlier Hellenic view of matter, a subordination to ideals of form and a religious-esthetic concept of unity became, in alchemy, mysticism and magic. Materials, the metals, were endowed with spirits and the concept of spirit in turn was both ideal (immaterial and *spiritual*) and material and substantial. This was not an unreasonable conclusion for an age that observed the transmutation of breath into fire or of candles into smoke. What had been the *form* of the metal, its color, its gravity, its sectility, became a kind of animate soul, susceptible to charms as well as to physical and chemical processes.

Out of this quest for transmutation grew the concept of the elixir or philosopher's stone. Originally the philosopher's stone was a mineral substance to be added to the smelter's charge to transform it to gold, just as ores of arsenic added to molten copper give it the appearance of silver. From an ore ingredient which would convert a base metal charge to gold or silver, the philosopher's stone became in time medicinal, a panacea for all ailments, by extension conferring immortality and even wisdom.

In the twentieth century, when the transmutation of elements is a commonplace concept of our chemistry, we can perhaps better appreciate what it was that the alchemists did and did not try to do. In the first place their concept of element was nothing like our own. "Of water there are two primary divisions, the liquid and the fusible kinds," says Plato in the *Timaeus*. Water meant not only our water but anything which could become a nonviscous liquid. A single word, *chalcos,* served the Greeks for copper, brass, and bronze. The Latins used the word *aes*. ". . . Every white metal and alloy fusible and alterable by fire was called originally lead," wrote Berthelot. Just as a modern handbook of steels may list thousands of alloys, all of which we commonly call steel, the ancients were aware that their bronzes, leads, and irons were mixtures. They prepared them by smelting a variety of minerals, and they understood how the kinds and proportions of the admixed minerals affected the final metal. What was lacking was our modern concept of the pure metallic element. The preparation of a hard, brassy *aes* required the highly colored carbonate ores of copper or the greenish-bronze sulfide, chalcopyrite. This had to be mixed with charcoal to reduce the ore, and with a salt such as soda for flux. Quantities of tinstone or of earthy white carbonate and silicate zinc minerals would greatly increase the yield.

When we remember that they were ignorant of metallic zinc (its volatility at low temperatures prevented the ancients from isolating it), hardly understood the nature of the fluxing process, and were completely without the concept of chemical combination and reduction, the alchemists' basic assumption of the alteration of metal by suitable stony ores is not only reasonable but highly practical. Their strange recipes—expose it to the sun with urine from an infant at breast,* select a young girl remarkable for her beauty and place her entirely nude below a spring of quicksilver in order that it shall be enamored by the beauty of the young girl—often conceal perfectly practical procedures, ill understood by their practitioners and fancifully distorted by the imaginations of the copiers of manuscripts. Urine was a commonly used source of salts in fluxing, tanning, and dyeing operations. The specification of infant's urine was a bit of conceit not unlike the thiamine, riboflavin, and niacin which saturate the modern hamburger bun. As for the quicksilver spring, "in a place in the far west where tin is found," what pagan revels of the ancient Britons of Cornwall were here being distorted by a fifteenth-century Syrian in the name of a third-century Alexandrian, Zosimos the Panopolitan?

The purpose of science is twofold. It is to lighten "the labors of men's hands," as Hooke put it in the seventeenth century, and also for the "pleasure of contemplative minds." It is hard to resist the thought that the aims of alchemy were ignoble. The alchemist began with the fraudulent objective of counterfeiting gold. He aimed by craft to secure for himself a dispro-

* Excerpts from translations by Berthelot, the late-nineteenth-century historian of alchemy, quoted in Stillman.

portionate share of the labors of his fellows. In desperation the Emperor Diocletian, about 290, ordered that all books on alchemy and metallurgy be burnt lest the currency and therefore the security of the state be debased. The alchemist's methods included reason and trial and error but the alchemist also appealed to magic. He literally sought out false gods and made sacrifice to them.

> Having received these ideas from our master . . . But our master dying before we were initiated and . . . we were told it would be necessary to evoke him from Hades . . . After invoking him several times . . . he replied that it was difficult to speak without permission of the daemon, and pronounced these words only —"The books are in the temple."*

This is from a manuscript as recent as the tenth or eleventh century A.D.! By contrast, there is an open-air quality to the Hellenic medicine of the fifth century B.C., an atmosphere of sunlight and exercise and simple diet. Hippocrates of Cos (c. 460–377 B.C.), following an Egyptian tradition going back into neolithic times, observed suffering people, described the symptoms of disease accurately and dispassionately, and left an ethical rule of disinterested devotion to the sick. How far ahead of the potions and charms and elixirs of medieval alchemy! Finally, immortality, which was to be won by the ancients for genius (Homer), or for courage as at Thermopylae, or for the faithfulness of a Penelope, the alchemists wished to coerce from the darker gods with the magical stone.

The mysticism of the alchemists was further complicated by astrology. Formerly each of the planets had been identified with one of the pagan gods and therefore men born under the sign of the appropriate planet shared something of the character of the god-planet, or were influenced by its motions and course. Beginning with an identification of gold with the sun and silver with the moon, a sympathetic identification of each of the metals with each of the planets was assumed. Saturn, with a period of revolution of thirty years, and therefore the slowest and most sluggish of the planets, was identified with

* Stillman, p. 156.

the heavy metal lead, and Mercury, with a period of eighty-eight days, with the fluid metal quicksilver.*

"Our jovial star reigned at his birth," says Shakespeare.

Persons born under the influence of Mercury were apt to be flighty, temperamental, quick in all ways—mercurial in disposition. The four humors of the body—blood, phlegm, black bile (*melancholy*), and yellow bile—were the determinants of character, and in turn were influenced by the supposed character of the planets.

This belief in the influence of the planets was a logical outgrowth of the concept of the universe as an organism. It became part of the apparatus of belief, part of the intellectual baggage of man, taking its place beside his theology, with its contradictions hardly challenged. Western theology taught that man was the center of the universe, the supreme object of creation, for whom and for whose moral dilemma the sun, moon, earth, and all living things had been called into existence. The greater world of the firmament with its stars and planets, the sun and moon, carried nightly about the earth, was the *macrocosm*. Man, the single point about whom and for whom all this turned, mirrored in himself the whole. Man was the *microcosm*. As the sun with its warmth and its seasons could reach across the intervening space and affect the affairs of men, so all the planets were believed to have their influences. Much later, when Newton explained the origin of tides in terms of the pull of the moon across intervening space, he was denounced for reintroducing to science the occult explanations of astrology.

If from a modern point of view the assumptions of the alchemists and astrologers appear naïve, they were nevertheless powerful stimuli to experimentation. In vain hope of boundless wealth and power, of evading judgment and death—a great body of practical discovery came into being, and, most important, passed into writing and therefore into permanent form.

* Jupiter was identified with tin, Mars with iron (possibly because of its rusty color), Venus with copper. See C. Leonardi, 1516, *Speculum Lapidum* (*The Mirror of Stone, or, The Seven Sympathies of the Metals, the Gems, and the Planets,* 1716) (Venice: M. Sessa).

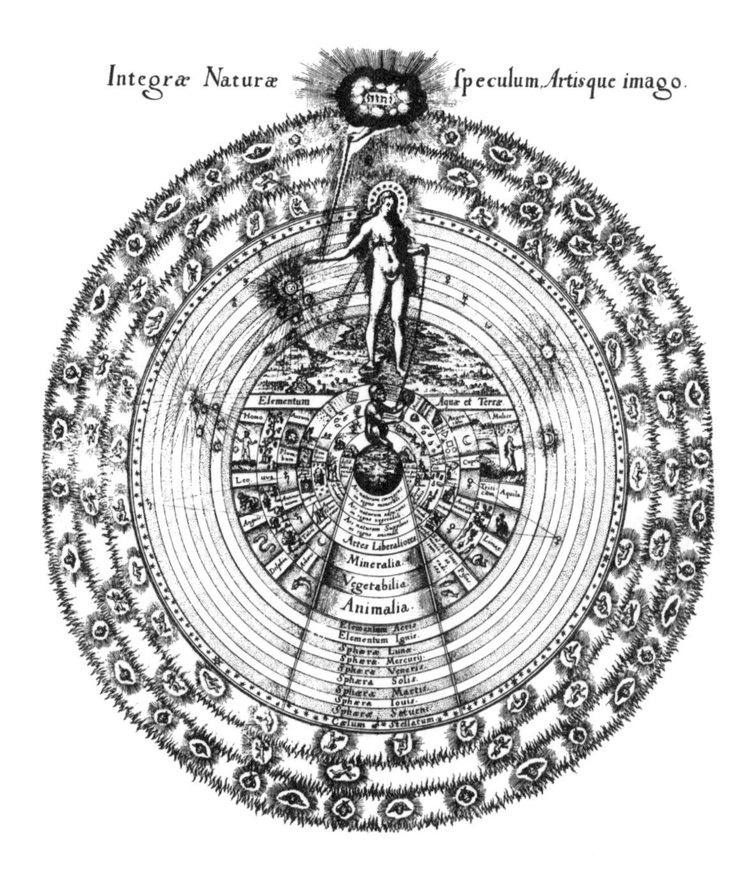

Figure 2. The Universe on the Eve of the Seventeenth-Century Scientific Revolution

Robert Fludd portrays the unity of God, Nature, and man. The Cosmos is a superbly ordered structure in which each part is related to the whole: the elements, the planets, the arts and crafts, and the stars. Spirit and material are complementary aspects of Pythagorean unity. (*Courtesy Basel Library.*)

In a search for the *alkahest,* the universal solvent or menstruum which would dissolve all bodies, a practical technology of acids and alkalis was well developed in classical times. The counter-search for the bottle in which to contain the alkahest led to extensive knowledge of chemical action and refractories. The processes of distillation and sublimation, bleaching, dyeing, etching, were introduced. In the cities of the greater Roman Empire of the first centuries of the Christian era, such as Alexandria in Egypt, factories produced glassware, cosmetics, and what we would call industrial chemicals, including limes and sulfates. Foundries and armories supplied tools and arms for the civilized world.

Gordon Childe has pointed out that the exports of the Mediterranean center were not only the products of these factories but, in time, the factories and technologies themselves, so that, beginning in the first century A.D., Spain, France, the Rhineland, Roman North Africa, and the Near East became independent centers of productivity. The civilization which spread was so materially advanced as to engulf even the barbarians who overran the empire. Not only were Roman techniques adopted but also Roman law, administration, language, and religion. The parallel to our modern competition in technical aid to underdeveloped countries is clear.

The decline of the Roman Empire was obvious by the fifth century. It was brought on by the rise of Christianity (according to Gibbon), or the rise of Islam (according to Pirenne), by an uncontrolled growth of rapacity and greed, by barbarian pressures, by the inexorable laws of history (according to the scientific historians), by social and moral failures, perhaps by the spread of malaria—for all of these reasons and perhaps in spite of these reasons, civilization in the West gradually faded. Not to quibble about whether Roman civilization was good, it is a fact that geese were pastured on the Capitoline hill after the sack of Rome by the Vandals in the fifth century.

The cities became villages. Society was fractionated. The size of the physical and political groupings into which men organized themselves became smaller. Gradually the city walls eroded, the treasuries were looted. The harbors filled with silt.

The aqueducts were broken and never repaired. Large numbers of people were no longer able to maintain themselves in the cities, and the unit of human society in the West became the farm-estate, the small village grouped around a central castle for protection. The population was static, small enough to survive on the fruits of primitive agriculture; trade was reduced almost to the vanishing point. The great manufactories of the empire shrank to cottage crafts and village smithies.

Through the Mediterranean–Near East area there was an infusion of fresh blood. The nomad peoples of the Arabian peninsula, stirred by the backwash of the dying Roman civilization, evolved a highly militant religion from Judaic monotheism and in the seventh century abruptly burst from the deserts into the Near East in a great conquering movement. The momentum of their surge was to carry their language and their faith around the Mediterranean, over Spain to the Pyrenees, through Provence, Sicily, much of Italy, down into Africa, over the whole of Persia and east into Central Asia and India. To the decadent Roman civilization they brought the Puritanism of a new faith.

In the West the triumph of Christianity had meant the destruction of much of pagan learning. The Serapeum of Alexandria, the great hospital-temple adjunct of the Museum, was destroyed and its manuscripts were scattered in 389 by order of the Emperor Theodosius, who was determined to stamp out all vestiges of paganism. The Museum itself was burned in 415 in the riots incited by St. Cyril, Bishop of Alexandria. Hypatia, Theon's beautiful daughter,* a mathematician, philosopher, and the ornament of the Museum, was dragged into a Christian church, stripped, and torn to pieces. The Academy at Athens, founded by Plato, was finally closed by order of Justinian in 529 and its scholars dispersed—mostly eastward, there to be overtaken by the conquering Arabs in a century or two. The dissident Christian sects—the Arians, the Nestorians, the Eutychians—and the Jews and the pagans dispersed from the Roman and Byzantine centers beyond the

* We do not really know what she looked like.

ends of the old empire, carrying with them the classical traditions and many of the manuscripts.

Outcasts, refugee scholars, and men in retreat from civil disorders banded together in small communities such as the Vivarium of Cassiodorus, c. 490–585. Early in the sixth century St. Benedict, founder of the monastery at Monte Cassino and of the order which bears his name, formulated the rule of obedience, labor, and devotional reading which from then on characterized the monastic movement. Here the manuscripts were preserved and copied through periods which saw the loss of literacy in the outside world. Even the Emperor Charlemagne was illiterate, his biographer Einhard wrote in the ninth century:

> He also tried to write and used to keep tablets and blanks in bed under his pillow, that at leisure hours he might accustom his hand to form the letters; however as he did not begin his efforts in due season, but late in life, they met with ill success . . . While at table, he listened to reading or music . . .

Isidore, Archbishop of Seville from 600 to 636, attempted to gather all of learning into one great *summa,* the *Etymologia sive origine,* combining the sacred works of the Church Fathers with the literature of antiquity. The *Origins of Words* was the first encyclopedia, the first attempt at a general synthesis preserving Aristotle and the pre-Christian learning. Translated into the vernacular and slowly spread through Europe, it could not prevent the decay of civilization, but provided inspiration and a basis for the revival of learning in the West when six centuries later the Gothic encyclopedists rediscovered Aristotle.

At first the conquering Arabs were fanatically intolerant, but their rapid success soon placed them in command of all that was best in the classical world. Briefly a brilliant civilization flowered. Observatories and universities sprang up for the study of Plato and Aristotle in Arabic translation. At the court of Bagdad the mathematician-poet Omar Khayyám echoed the Preacher of Ecclesiastes in exquisite verse. From Russia came slaves and furs and honey, bought from the Greeks of Byzantium with Indian peacocks and Chinese silks, and scimitars of Spanish steel, their scabbards encrusted with gold and gems.

While in the West the Roman roads fell into ruin and the Greek marbles were burned by the peasants for manure, in the East, by the tenth century, across an empire that spread from the Pyrenees to the Indus, Persian star worshipers, Jews, Christians of all sects, Buddhists, neo-Platonic mystics, and Hindus mingled freely, translating Archimedes and assimilating the mathematics of India.* Briefly the world was again secure and civilized, with trading dhows voyaging east to India, west to the Gates of Hercules, and from there north to the shores of France.

It is in Arabic form that the foundations of science reached the West, and we preserve the record of this transfer and of the orientation of Arabic science in a multitude of Arabic terms—chemical: alcohol, alkali, niter, alkahest, alembic; astronomical: almanac, zenith, nadir, the names of the visible stars (Aldebaran, Algol, Marfak, Betelgeuse); mathematical: algorithm, arithmetic, algebra. The stigmas of paganism, heresy, and Mohammedanism, as well as the dubious association with the black arts of alchemy and astrology, accompanied the slow filtration of science to the West.

The Arab power did not last long. By the eleventh century it was crumbling under the attack of the Turks in the East, themselves soon to succumb to the Mongols. The Arabs were also in retreat in Mediterranean Europe. Gradually, in the same way that their conquest of the classical world had stimulated the Arabs to civilization, the conflicts of Europeans with Arabs in the eleventh and twelfth centuries stimulated a revival of interest in learning. Now the works which had been translated from Greek into Arabic were translated into Latin.† By

* Al-Razi said, "A Chinese scholar came to my house and remained in the town [Bagdad?] about a year. . . . In five months he learned to speak and write Arabic . . . When he decided to return to his country he said to me . . . 'I am about to leave. I would be very glad if someone would dictate to me the sixteen books of Galen before I go.' " Al-Nadim (d. 995), translated by Ferrand, quoted in Needham, vol. 1, p. 219.

† Ptolemy's *Almagest,* Euclid's *Elements,* Galen, and Hippocrates were translated from Arabic into Latin by Gerard of Cremona (1114–1187). The

the thirteenth century Frederick II, the wonder of the Western world, presided over a center of translation and learning at Palermo in Sicily, bringing Arabic astronomy, alchemy, and mathematics into Latin culture. Perhaps its single most significant achievement was the *Liber Abaci* of Leonardo of Pisa, which introduced Arabic numerals, and therefore modern arithmetic, to the West. Of scarcely less significance, the new astronomical tables of Alfonso X of Castile (d. 1284) opened the way for the age of navigation.

The term alchemy itself is of Arabic origin, based on the Greek term *chemeia* which from about the fourth century referred to the art of metalworking and particularly to metallurgy, including transmutation of gold and silver. Zosimos, the famous chemist of Alexandria, wrote at the end of the third century that the sacred art was brought to man by angels who lost their place in heaven through their love for mortal women. Their secrets were supposedly found in the book of *Chemes* or *Chymes.* Modern scholars prefer a more prosaic explanation and attribute the word to the old Greek *khem,* meaning black, the Greek name for Egypt (believed to refer to the fertile soil of the Nile delta).* From its inception, therefore, chemistry was the black art, arising in the Black Land. Tradition had it that the first alchemist, and the source of all the secrets which were to be discovered, was the god Hermes Trismegistus, Hermes Thrice-greatest, who in the Roman pantheon was merely the messenger of the gods, but in Egypt was Thoth, the Ibis-headed source of writing and philosophy. Adepts were called Hermetics, and even today we speak of a hermetic seal—faint relic of a time when the winged insignia of Hermes were pressed into the wax closing an alchemist's retort to protect it from the meddling of the uninitiate. Moses, his sister Miriam, and an early Egyptian priestess conveniently named Cleopatra were other supposed founders of the divine art.

---

astronomical tables of al-Khwarizmi (c. 850) were translated by Adelard of Bath (1090–1150), his *Algebra* by Robert of Chester.

* A still more prosaic explanation attributes the name to the Greek *cheo,* a word linked to metal founding.

In the first few centuries of the modern era, science, which in ancient Egypt had been in the hands of the priesthood, in Egyptian Alexandria centered on alchemy. If Rome was the capital of the ancient world, Alexandria was its crossroads. It was a city of religious ferment, practicing the cult of Mithras, brought from Persia; the worship of Serapis, introduced by the Ptolemies and identified by the Egyptians with Osiris-Apis, the Bull-headed god; the worship of Isis and Horus; the worship of the Roman Pantheon; Judaism with its various Hebrew messianic sects; the Eastern mystery cults with their secret rites and orgies; the cult of the Phrygian Cybele; the cult of the Mater Magna or Great Mother; and the dissident Christian sects—Arians, Nestorians, Eutychians, Catholics, and Gnostics. In this environment alchemy too had much of the character of a secret society. Many of the early alchemists were Gnostics, which is to say members of a neo-Platonic cult combining Christian, Jewish, Chaldean, and Egyptian mysticism. The early Greeks had despised the practical arts, reserving them for slaves. In Hellenistic Alexandria of the first centuries, the Greek ideal fused with technology and produced alchemy, a source now not only of dangerous ideas but of wealth and power. The alchemists adopted secrecy and mystery in self-defense. They failed to produce a consistent and rational literature, but they did begin to deal with metallurgical, chemical, and medicinal problems.

Few of the alchemists enter into the history of science. The Leyden Papyrus of the third century lists Plato, Aristotle, Hermes, John the Archpriest, Democritus, Zosimos, Olympiodorus, Stephanus the philosopher, Sophar the Persian, Synesius, Dioscorus, the priest of the Serapion at Alexandria, Ostanes the Egyptian, Comarius the Egyptian, Marie, Queen Cleopatra, Porphyrius, Epibechius, Pelagios, Agathodemon, the Emperor Heraclius, Theophrastus, Archelaus, Petesius (Isidore), Claudius, "the anonymous philosopher," the philosopher Menos, Pauserius, and Sergius, as the masters "celebrated above all," the makers of gold. We know of them through manuscript copies usually much later than the dates of the originals, through references in other works, and in rare cases, such as that of Mary, through their discoveries (hers was the bain-marie or water bath). The first of the line was Democritus, who in the later tradition became confused with Democritus of Abdera (420 B.C.), who founded the atomic philosophy. Until well into modern times it was common practice for authors and commentators to ascribe their works to famous figures of the past. Medieval manuscripts were ascribed to Hermes, to Moses, to Aristotle and Democritus, later to the Church Fathers and the great scholastics. The alchemical Democritus and his successor Zosimos were early Alexandrian neo-Platonists whose wide knowledge of metallurgical practice and chemical process was not equaled by their successors. The alchemical tradition began with a direct acquaintance with experience, but later degenerated into a literary exercise.

After the seventh century Alexandria was Arabicized. The first of the Arabic sources in alchemy was Khalid, or Khaled ibn Yazid (635–704). The most famous Arabian alchemist was Jabir ibn Haiyan (Djaber, Geber, c. 776). He is well known because a Spanish alchemist (c. 1300) wrote a number of works in Latin under this name, thus greatly increasing the corpus of Geberian alchemical works known to the Renaissance. It is possible that even the original Geber was an apocryphal creation of the Brethren of Purity, a society of Basra (now in Iraq) which existed about 975–1000.

The greatest of the Islamic alchemists was the Persian al-Razi (850–925), known as Rhazes. Physician and alchemist, he was reputed to have written more than a hundred books. One of them, the *Book of Secrets,* was an encyclopedic guide to alchemy with clear and detailed descriptions of reactions and apparatus and a classification of chemical substances. In this way a wholly empirical tradition of chemical description and classification began side by side with the principally intellectual and logical doctrine of the elements. Ibn Sina (Avicenna of Tadzikistan, 980–1037), successor to Rhazes as first physician to Islam, tried to separate medicine from the less savory aspects of alchemy and doubted the possibility of transmutation. Both are important for their influence on the medieval scholastics, and their names were borrowed by later mystics.

A major source of Central Asian alchemy was the works of the Brethren of Purity or Faithful Brethren. Their writings gave particular emphasis to an ancient suggestion that metals were a mixture of mercury and sulphur—the fusible watery element of Plato and the active, male, fiery sulphur. They developed a consistent scheme of the origins of ores.

> The various waters which mingle in the interior of the earth are by heat volatilized to the upper strata in crevices and cavities there becoming condensed and thickened by cooling, and again percolating downward, mix with earthy particles and by the heat of the earth are changed to quicksilver.*

Sulphur was produced from air and earthy particles by the heat of the earth. Sulphur combined with mercury in different proportions to produce the metals, according to the length of time and the temperature of their combination. The sulphur-and-mercury theory of metals was derived from metallurgical practice. A large number of metallic ore minerals—the principal ores of copper, lead, zinc; the metallic minerals associated with tin, gold, silver; the arsenides, antimonides, and sulfo-salts—yield sulphurous fumes on roasting. The reduction of cinnabar (HgS) to mercury and sulphur and the converse synthesis of cinnabar from mercury and sulphur were well known to alchemy. In the early Renaissance this idea would be transformed into the three-principle idea of mercury, sulphur, and salt.

The peculiarly medieval cast of thought which goes by the designation of scholasticism began with Peter Abelard (1079–1142), famous for his ill-fated romance with Héloïse. Abelard's brilliant revival of the dialectic and its employment in religious discussion attracted large numbers of students and led to the rise of the universities. In a sense this was the invention of theology—literally God-logic, or the science of God. Aristotle and the Aristotelian logic which had derived from Zeno and Parmenides, via Socrates and Plato, furnished the form in which the scholastics cast their religion.

In the great age of the scholastics, the thirteenth century,

* Stillman, p. 211.

Albertus Magnus (c. 1206–1280) and his pupil Thomas Aquinas (c. 1225–1274) completed this marriage of reason and faith. The ideal structure of the Middle Ages, of a single community of mankind in the Church, under law grounded in nature and reason, was largely the work of St. Thomas in the *Summa Theologica.* A similar synthesis had been performed for Mohammedanism by Averroes (1126–1198) of Córdoba. The synthesis of Aristotle and neo-Platonic truth with Judaism and the codification of the Hebrew faith were accomplished at about the same time by Moses Maimonides (1135–1204), in Egypt.

This re-establishment of religion on logic and the buttressing of faith by reason, which we look upon today as the highest achievement of the Middle Ages, was not entirely acceptable to orthodoxy. Abelard was condemned twice for heresy. Maimonides too was attacked. The work of Aquinas was saved from condemnation only by the active defense of Albertus Magnus.

The revival of Aristotle entailed the introduction of a vast body of secular learning, all, in fact, of that pagan learning which had been nearly extirpated centuries before. The first of the neo-Aristotelian encyclopedists who attempted to summarize the whole of knowledge was Albertus Magnus. In his great encyclopedia he tried to write one work for every book of Aristotle's. Vincent of Beauvais (c. 1190–1264), Roger Bacon (c. 1214–1294), and Bartholomaeus Anglicus were the other great collators of the thirteenth century. They were in every case clerks, part of the system of monastic orders which we recall today only in the peculiar garb of academic ceremony. Albertus, Vincent, and Aquinas were Dominicans; Bacon and Anglicus were Franciscans (orders founded in 1215 and 1210 respectively).

The thirteenth-century encyclopedists not only brought together the Greek and Arabic sources but were themselves original contributors to knowledge. Albertus was a naturalist, classifying the vegetable and animal kingdoms. Bacon was the great prophet of science, setting himself the task of a complete reform of all knowledge on logical and experimental bases.

The rewards for this endeavor, he prophesied, would be complete control over nature—human flight, the circumnavigation of the globe, the prolongation of human life, the elimination of labor and want.

With the encyclopedists there was a conscious attempt to bring alchemy down to earth. The doubts concerning transmutation and the puritan attitude toward magic which already marked the Central Asian chemists (Avicenna in particular) were characteristic of the neo-Aristotelian rationalists. "Let the artisans of alchemy know," wrote Vincent of Beauvais,

> that it is not possible for species to be transmuted, but they can make things similar to these, as by tincturing white [metal] to a yellow color so that it may seem to be gold . . . they may produce in it such qualities that they may deceive men in it . . .

Still, the weight of all authority could not be entirely rejected.

> . . . it is true that by the ancient philosophers and by artisans in our time it [transmutation] has been proved to be true.*

The very fact of publication was a change in the fundamental character of alchemy, denying the mystery, denying the exclusiveness. The encyclopedists, particularly Roger Bacon, were concerned with the usefulness of chemistry, listing the recipes, classifying the ores and minerals, the glazes, the methods of preparation and refinement. They were technologists and artisans as well as metaphysicians. Much of their Aristotle was neo-Platonic apocrypha, and they were unable to distinguish the pseudo-Avicenna from the true, the pseudo-Rhazes from the original. In the same way later alchemists were to borrow the names of Albertus, Bacon, and Vincent of Beauvais.

The civilization of the Middle Ages, with its trade, its manufactures, its transportation, and its warfare, required at least the elements of industrial chemistry. Mining, which had been extensively developed by the Romans, died down for a time and Europe existed on metals which had been mined in the past. In 745 Schemnitz, the first of the Saxon silver mines, was

* Stillman, p. 246.

opened in the Carpathians. The industry spread until by the time of the Renaissance, Saxony became a world center of mining and mineralogy. The methods of the Saxon miners, faithfully recorded by Georg Bauer (1494–1555), a physician of Joachimsthal in the Erz Gebirge who wrote as Agricola, show few changes from the metallurgy of the classical era. According to R. J. Forbes,* some coal came into use as the growing shortage of timber made the charcoal required in smelting more expensive, but the coal was only an auxiliary in preliminary operations. The general shortage of labor in medieval Europe, as well as contacts with the Far East, promoted the development of machinery and the application of water wheels to the bellows and hammers of the medieval forge. In Spain the Moors discovered that copper was precipitated from the waters of sulphide mines by running it over iron. They developed an efficient and humane industry on this basis simply by running water through the old Roman mines. By the thirteenth century the mining and metallurgy of iron was being carried on extensively throughout Europe. Forbes points out that Richard I used 50,000 horseshoes forged in the Forest of Dean for his twelfth-century crusade.

Although medieval metallurgy used no processes that were unknown in classical times, there was a steady development of size and efficiency of furnaces for the reduction of iron ore. This led in the thirteenth century to the production of pig iron as the raw material for secondary processes, either for reducing the carbon content below about .3 percent to make wrought iron, or for increasing the carbon content above about .3 percent for the production of steel. With the development of tall *stück* and *blas*-ovens in the Rhineland, with drafts forced by water-driven bellows, it became possible to reach higher temperatures. Iron of any desired carbon content was produced by simply regulating the draft. For the first time cast iron (iron containing more than about 6 percent carbon) could be produced directly, rather than by alloying iron with expensive and embrittling arsenic, copper, or other metal.

Despite these practical accomplishments, the metallurgy of

* Singer, et al.

iron was still understood only in reverse, since neither the concept nor the techniques of carbon analysis were known. It was believed that carburization was a purifying process and that wrought iron (with the lowest carbon content) was the least pure, while cast iron (with the highest carbon content) was, finally, the pure metal.

The invention and exploitation of glass blowing probably began in Syria about the first century. The skills and techniques of the later Roman glass blowers were not inferior to those of Renaissance Venice, so that not until the nineteenth century do we discern chemical innovation. The stained-glass windows of the Gothic cathedrals represent architectural but not technological advance.

The development of ceramics, however, was greatly influenced by the Chinese invention of porcelain, a ceramic of special clay and feldspar, characterized by exceptional whiteness, translucency, and a metallic resonance. The porcelain technology, requiring particular minerals and high temperatures, could be imitated by the Arabs and in turn by the European ceramicists, but was not successfully duplicated until the eighteenth century. In China the techniques were developed from the fifth century on, and by the ninth century Tang ware was being exported as far as Egypt and Samara in Mesopotamia.

According to the historian Joseph Needham, in the early Middle Ages Jewish merchants known as the Rhadanites traveled regularly between China and Provence by way of Damascus, Oman, Hindustan, and Khazaria in the Crimea. To China they carried eunuchs, slaves, brocades, furs, and swords which they exchanged for musk, aloes, cinnamon, camphor, and herbs. During this period and slightly later, the magnetic compass, the power-driven metallurgical bellows, gunpowder, paper, movable type, and the techniques for casting iron, to name only a few, passed from China to the West.

Attempting to imitate porcelain, the Arabs developed the tin-glaze, a wide palette of glaze colors, and very elaborate techniques of slip (the application of clay slurry for special effects), all of which soon spread to the West. Here again, even the discovery of the true China porcelain process at Meissen in 1709, while leading to very fine ceramics indeed, never quite matched the Persian and Islamic imitations or the medieval Chinese originals.

In speaking of the high quality of the metallurgical, glass, and ceramic technologies of the past we are not simply acknowledging their esthetic superiority, overwhelming as this is, but also their purely technical superiority. Smothered as we are in a flood of commercial cajolery designed to persuade us that Dacron is superior to eiderdown, rayon to silk, and pressed luncheon meat to steak, we are scarcely aware of the dying out of irreplaceable skills. Few glass blowers practice today, and none are the heirs of generations of jealously guarded and cultivated techniques. The master Cordovan swordsmiths, the Tang potters, were geniuses of particular skills, perhaps the best among thousands, occupying positions in society which we reserve for professional ballplayers.

Soap is believed to have originated with the Germanic tribes, who brought it to the Roman world by the fourth century. In the Middle Ages it was a common article of manufacture and export. Hard soap made from olive oil and soda ash came from the Arabs, especially in Castille and Marseilles; soft soaps for the cleansing of cloth came from Scandinavia. Carbonates of soda called *nitrun* in Egypt and *natrun* by the Arabs were mined in desert regions, and also made by leaching the ashes of certain shrubs. Potash (lye) used for soft soaps was obtained by leaching ordinary ashes. Lime was mined, much of it in the form of marble looted from classical ruins, and roasted to quicklime.

Gunpowder or black powder is a mixture of saltpeter (KNO_3), charcoal, and sulphur in the ratio of seven or eight to one to one. Heat of friction or a blow or spark or flame starts the reaction involving the loss of oxygen from the niter and its combination with the other ingredients. This is accompanied by the release of a large amount of energy—an explosion. In modern parlance, there is a chain reaction, the spark setting off the reaction in the adjacent grains of powder which set off the grains adjacent to them and so on to the end, all of this proceeding so rapidly that the chemical energy of the reaction is released in a relatively short time in a relatively restricted space. If the powder is packed in a confined space such as a bamboo or cardboard tube, the result is an explosion. According to Needham, the properties of such saltpeter mixtures were first discovered in the Taoist temples of China toward the end of the Tang dynasty, about 900, a period which saw also the widespread dissemination of printed books in China.

Incendiary or chemical warfare in one form or another was of ancient origin. The use of burning tar, sulphur, pitch, saltpeter, boiling and burning oils, smokes, torches, and quicklime was universal. These were hurled or projected on the enemy, most effectively in fire pots. Liquid mixtures of flaming sulphur and naphtha (petroleum) in clay pots were hurled by catapults or other elastic engines, acquiring the name of Greek fire during the centuries when the Byzantines successfully used them in defense of Constantinople. These mixtures were sprayed on the enemy through huge bronze force pumps, but nothing like an explosion was produced nor apparently were any of these mixtures self-igniting.

By the year 1000 black-powder grenades and bombs were being used against invading northern Tartars by the Sung emperors. One would expect the enormous superiority of this new explosive weapon to have been decisive in warfare, particularly when it was placed at the disposal of a large, well-organized, and prosperous state. Perhaps indeed it did decide the first battles in which it was used. But the Sung were soon driven south by their enemies, who acquired gunpowder from them. Early in the thirteenth century the Mongols overran a Chinese munitions factory. In the same century a recipe for gunpowder appears in a text of the *Book of Fires for Burning Enemies* of Marcus Graecus, probably a contemporary addition to an ancient compilation; and Roger Bacon refers to gunpowder in works written around 1268. By the end of the century firearms—that is, weapons employing the explosive force of gunpowder as a propellant—were more advanced in Europe than in China. Any parallel which we attempt to draw with the invention of nuclear explosives is clearly depressing.

One additional discovery worth mentioning is the development of distillation processes capable of producing alcohol (Arabic: the essence or subtle part) in sufficient concentration to burn, a development originally attributed to Arabian alchemy but now believed to have originated in Italy about 1100. Using quicklime for taking up water and repeated distillation, a thirteenth-century alchemist who appropriated the name of Ramón Lull obtained absolute alcohol. Its solvent properties led to its use in cosmetics, drugs, and paints. Medicinal liqueurs like Benedictine, prepared in the herb gardens of the monasteries, were widely employed against the plague and other diseases. Gin and brandy were invented and with them came the problems of alcoholism and, in the thirteenth century, the first temperance laws.

Of the other chemical techniques of the Middle Ages, nitric acid, first described by the pseudo-Lull and a pseudo-Geber, was prepared by distillation of niter. It was used to dissolve silver out of gold and for etching.

The preparation of aqua regia by the addition of sal ammoniac to nitric acid was described by pseudo-Geber. This was the first substance known to dissolve gold. Sulphuric acid from the combustion of sulphur or the distillation of vitriolic salts; hydrochloric acid from the distillation of sea salt (NaCl); and vinegar and related acids are part of the technology of this later period. The techniques of dyeing, of preparing paints and pigments, varnishes and inks, reached and even surpassed the levels of quality of the ancients.

The use of lapis lazuli for the remarkably deep ultramarine of such treasures as the *Très Riches Heures* of the Duc de Berry is an example. Azurite, an inexpensive blue carbonate of copper, was frequently substituted for the precious lapis lazuli. Unfortunately it readily alters to the green carbonate of copper, accounting for the peculiar penchant of late medieval artists for green-eyed subjects against a greenish sky. With the development of glazes under the influence of the Chinese by way of Islam, less expensive if less satisfactory pigments could be prepared. Dyes made from crushed insects (vermilion, crimson) and purple dyes from the secretions of shellfish had been in use since classical times, as well as plants cultivated especially for their use in dye manufacture—madder, indigo, woad (blue), and safflower (yellow). Alum, aluminum sulfate, was used as *mordant* to prepare fibers for a fast dye.

Papermaking, with its accompanying development of inks, particularly after the spread of printing, swells the volume of industrial chemical activity of the early Renaissance. Paper was invented by Tshai Lun (d. 114) in the latter part of the Han dynasty. When we consider that a parchment folio of two hundred sheets required twenty-five sheepskins, the significance of the discovery is obvious. Linen rags were beaten to pulp and mixed with water to form a kind of slurry. This was caught on a wire screen, drained, and the fibers felted, then dried. Earlier Han manuscripts had been carved in wooden tablets or written on silk, but the idea of printing from these tablets on the new paper was not immediately grasped. The technology of papermaking and the chemistry of inks were to develop over the next five hundred years.

Needham suggests that it was the spread of Buddhism, with its interminable repetitions of names and prayers, which led to the invention of printing in the Tang dynasty. Certainly the use of seals, etching, and engraving, and even the rubbings of engraved stone or metal monuments were known from early times. By the end of the Tang, Taoist books were widespread in Szechuan, and blocks for the printing of the Confucian classics were cut in 932–953. Movable type, introduced in the Sung dynasty (960–1279), probably from Korea, was by no means a simple extension of the basic idea of printing, since it required the complicated technology of interchangeable parts (the letters). This in turn entailed the metallurgy of type founding.

Paper reached Europe first about the twelfth century, after the technology had spread through India and Islam. By the end of the thirteenth century there were Western references to block printing of religious pictures, playing cards, and paper money—surely a significant juxtaposition. The paper money of the Mongol Empire was described by Marco Polo on his return to Venice. Europe adopted the block printing of religious pictures and playing cards. The ordinary inks, lampblack (mentioned by Pliny), or plant galls treated with ferrous sulphate in gum or glue solutions, were used at first. By the time of Gutenberg (c. 1400–1467) printer's ink of lampblack in a varnish prepared from boiled linseed oil was developed. It seems probable that the first type founders, such as Gutenberg —who was a goldsmith by trade—adapted the jeweler's and pewterer's techniques of precision casting.

The work of the first printers was of a quality equal to that of the finest Renaissance draftsmen. In Gutenberg's Bibles each letter is a small masterpiece of precision and design, each line characterized by perfect alignment and even register, each page by elegance of composition and clarity of impression.

As we approach the Renaissance and particularly the scientific Renaissance, the figures of Copernicus, Brahe, Kepler, and Galileo emerge and the world becomes more familiar. The Americas are discovered and immediately settled; the nations of Europe appear in much their present form. Linguistic, religious, and political differences no longer involve the alien concepts of Ostrogoth and Visigoth, Arian and Eutychian, but those of Spain or France, of Protestant and Catholic. We recognize the intellectual world in which they moved, these ancestors of ours. We still read their Shakespeare and King James Bible, while we reserve St. Thomas for scholars and theologians, Chaucer for translation.

What is it that caused this Renaissance, this age which proudly claimed for itself the mantle of antiquity? Was it the fall of Constantinople to the Turks, so that the Greek scholars were driven west to Italy, bringing with them the Greek and Latin literatures? Was it the discoveries, the circumnavigation of Africa, the voyages of Columbus and Magellan? Was it the introduction of printing to the West, the spread of literacy, the rise of Protestantism? It is true that on close examination no real dividing line may be drawn which would enable us to say, "Here, at this place (Italy? Flanders?) at this date (Dante's *Commedia?* 1492?) the past ended, and the modern world began."

Yet the forest does thin out as it merges into the plain, and if we cannot draw a line of demarcation, nevertheless we are well aware of when we are in the Middle Ages, of when we have crossed the border into the modern world. Boyle, in the late seventeenth century, even Newton, still hopefully search for transmutation. But Boyle's book is entitled *The Sceptical Chymist* and Newton's *Principia Mathematica.* We have indeed crossed over a divide.

The concept of elements (going back into prehistoric time), the world as animate, as organism, the identification of man with universe, of planet with metal, the confusion of spirit or mind with substance (air, vapor), the magical attributes of number and form, these paraphernalia, this intellectual baggage which the migrant European man carried lovingly across the frontier to the modern world, have not been entirely discarded even today, but were soon ignored, relegated to the attic or to museums presided over by cranks. Other ideas, whose influence was not immediately discernible, led to the great revolution in human existence which we may appropriately designate as the Age of Science, a relatively recent subdivision of the Age of Letters, and one may speculate as to whether or not an inevitable result.

From the past of classical civilizations, the dark ages of their decline, and the long period of European maturation, the modern world drew for its age of science a sense of unity and continuity. Despite the wars and schisms, the many exceptions, the indefinite boundaries in time and space, a single culture had evolved—a European society, drawn from rationality and the Judaic invention of God-morality by the classical method of the dialectic. It was the product of that scholastic synthesis of Aquinas but also of the centuries of absorption first of Roman law, then of Church law, but always of the concept of a world of law, a world of purpose and order under God, and of natural law and reason and number. This too the migrants to the New World carried in their baggage but carried unaware, unconscious of its presence, stowed away in the midst of the more obvious trappings of alchemy and astrology.

Finally we observe the revival of trade, the world increase in population, and the rise of the middle class. Technologically the achievements of the past had been largely recovered,* even perhaps surpassed. In place of the artisan slaves of the Roman and Alexandrian factories the guilds of master craftsmen appeared. Utopia had not yet been reached, as the revolts of the weavers in Florence and the Jacquerie in France attest; but from the standpoint of technological advance, the transformation of those responsible for the actual operations with materials, from slaves into an accepted class of society, into Christians, and especially into the powerful bourgeoisie of such cities as Bruges, Lille, London, Florence, Zurich, and the cities of the Hanseatic League, made it possible for science (philosophy) to merge with science (technology), without which interplay of theory and experiment little change would be possible.

Side by side with the elements and principles, the Renaissance inherited from alchemy and industry a knowledge of processes and the behavior of materials, of the classification of minerals and substances, techniques of analysis and a pharmacopoeia. There lacked only the conceptual framework for a science of chemistry to appear. The strides which were being made in physics in the chain of ideas that led from Copernicus to Newton made the requirements for a science of chemistry clear and its achievement inevitable. The relative complexity of the task—there were after all only five planets from which the cosmologists fashioned a mechanics, but more than a hundred elements enter into the material world—insured that the development of chemistry would be slow. Mechanical ideas like atomism are easily conceived, but they are brought into the numerical agreement with experience that is characteristic of modern science only by highly sophisticated experiments such as X-ray diffraction or Millikan's oil-drop experiment.

None of this could be foreseen by that generation in the seventeenth century who were making the transition from medieval alchemy to modern chemistry, for whom mathe-

* Let us raise one demurrer for the Imperial or Tyrian purple, the dye of the Roman emperors, often imitated but according to tradition never equaled, its secret lost forever with the fall of Constantinople.

matics, planetary motion, light, metallurgy, geology, transmutation, and the nature of the air were alike mysterious problems, no one of them offering easier prospects of solution than any other.

Here at this stage in history appears the first distinction between the "hard," the "positive" sciences with their confident answers, and the "soft" sciences, barely able to describe the limits of their problems. For two hundred years chemistry was to limp behind, with intuition struggling to shed the double handicap of its heritage from alchemy and the fixation of successful Newtonian mechanics.

4

The Scientific Renaissance

WE LIVE SUBMERGED AT THE BOTTOM OF A SEA OF ELEMENTARY AIR, WHICH IS KNOWN BY INCONTESTABLE EXPERIMENTS TO HAVE WEIGHT.

—TORRICELLI, 1644

Ctesibius of Alexandria, about the second century B.C., is said to have discovered the elasticity of the air and to have invented the suction or piston pump which was widely used in the ensuing centuries and became a necessity for the later medieval mining industry. As ores near the surface were exhausted miners were forced to work deeper and deeper underground. But both crevices in the rock and surface openings pass ground water, and the mines soon filled unless they were kept pumped out. As the mines deepened the task became more and more difficult. With the hollowed logs and other crude pipes used, the suction or lift pumps would not operate much below ten feet.

As we know now, such a pump will not operate at all with a lift over a limit of thirty-two feet. Aristotelian philosophers explained the limited action of the suction pump with Aristotle's dictum that nature abhorred a vacuum. Lifting the piston in a chamber with air excluded would produce a vacuum unless water rose in the space behind the piston. Apparently, however, natural abhorrence of vacuum did not extend beyond thirty-two feet.

When in the seventeenth century Galileo Galilei, who had already rocked the Aristotelian conceptions of motion and of the solar system, attacked the problem, he concluded erroneously that a column of water had a cohesive strength that would hold it together. But if he had failed in this respect, he had succeeded in another. He invented the air thermometer, which consisted of a glass bulb at the end of a long tube. The tube was inverted with one end immersed in a dish of mercury. As the air cooled and contracted, mercury rose in the tube. As the air warmed, it expanded and forced down the level of the mercury in the tube.

Galileo was condemned by the Inquisition for his support of the Copernican idea of the solar system, but after some years he was allowed to live in a kind of house arrest at Arcetri, in the hills above Florence. Here, gone blind, he was joined three months before his death by his successor as court mathematician for the Grand Duke of Tuscany, Evangelista Torricelli (1608–1647). After Galileo's death Torricelli and Galileo's pupil Viviani continued his experimentation, and soon found that it was the weight of the air, the fourteen pounds per square inch with which the atmosphere presses upon us, that forces the water thirty-two feet up after the piston in a pump, or holds up the 2.4-foot mercury column in a barometer (mercury is 13.6 times as dense as water). The partial vacuum left above columns of mercury in tubes greater than twenty-nine inches, or 2.4 feet, in height became known as the Torricellian vacuum.

These experiments were performed at Florence about 1643 and were a topic of discussion at the London meetings of the mathematician John Wallis and his friends, meetings which led to the founding of the Royal Society of London in 1662. Torricelli's ideas were confirmed by the ingenious experiments of Blaise Pascal (1623–1662), who is famous not only as a mathematician and the constructor of a computing machine, but as a theologian in the Jansenist movement within the French Church. In 1648 Pascal, whose chronic ill health had contributed to his conversion to the mystic and evangelic cause of Jansenism, had his brother-in-law, Florier Périer, take a barometer to the top of the Puy de Dôme, reasoning that the weight of the air would be less there and therefore the height of the column of mercury would drop. The success of the ex-

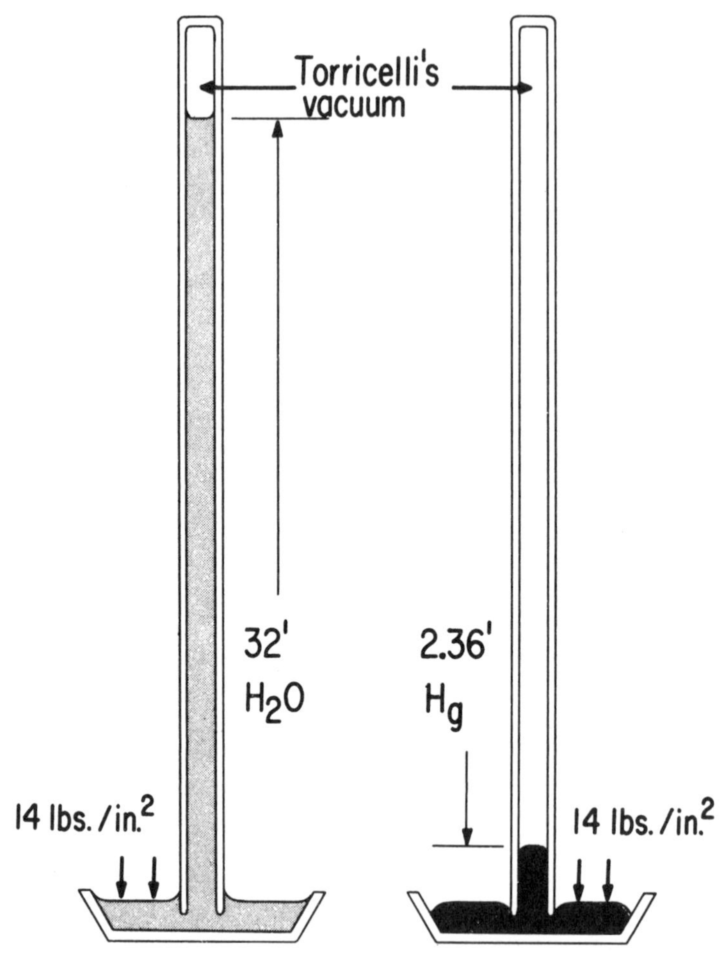

Figure 3. Torricelli's Barometer
The pressure of the atmosphere supports a column of water 32 feet high or a column of mercury about 1/14th as high. This occurs because mercury is about 14 times as heavy as an equal volume of water.

periment dramatically established the Torricellian concept, and it may be considered one of the first significant experiments of the science of geophysics.

At this time, possibly inspired by the experiments of Torricelli and Viviani, Otto von Guericke (1602–1686), burgomaster of Magdeburg, in Germany, used a piston in a closed cylinder to duplicate the Torricellian vacuum and thus invented the air pump. It was von Guericke who dramatized the new philosophy by staging the experiment of the Magdeburg spheres before the Imperial Diet at Ratisbon. Two small bronze hemispheres were fitted together and the air within was pumped out. Two teams of eight horses each could not pull the spheres apart. But when the valves were opened, breaking the vacuum, they fell apart of their own weight.

Robert Boyle read of von Guericke's experiment in 1657, when he was setting up a laboratory at Oxford. He engaged Robert Hooke (1635–1703) as his assistant, and Hooke soon designed and constructed an improved pump with which they carried out those experiments on the corporeality of the air so fascinating for the new scientists and so incomprehensible to the nonscientists, who could think of nothing more futile than "to weigh Aire." But it was these experiments, as well as his atomistic and Baconian preconceptions, that led Boyle to kinetic-molecular hypotheses which, by way of Bernouilli and Joule, flowered in the modern kinetic-molecular theory. The experiments also led Hooke to foresee the steam engine and, through Papin and Newcomen, to bring it into being, and with it the age of power. Equally important, the attention paid to the corporeality of the air turned chemistry toward the concepts of gas and of the different physical states of matter. They led to that flood of experimentation which found differences among gases and established the laws of gases, and also to Lavoisier's principle of the conservation of matter, in which chemistry as we know it today was born.

Robert Boyle (1627–1691), the youngest son of that Elizabethan daredevil the Earl of Cork, wrote to a friend in 1646 describing ". . . our new philosophical college, that values no knowledge, but as it hath a tendency to use." The Puritan revolution was in full sway. Young Boyle on an intellectual plane reflected the mood of an age which sent the Pilgrims to the New England wilderness* and the Roundheads, singing psalms, against the Cavaliers of Charles I. The prophet of the new philosophy was Francis Bacon, its origin his *Novum Organum* (1620), which he had written as the Elizabethan answer to the *Organon* of Aristotle; its method was inductive, rather than deductive as with the scholastics. Finally, Bacon promised, the new philosophy would lead to the *Instauratio Magna,* the restoration of mankind through science to the earthly paradise before the Fall.

The inductive method, as Bacon's followers understood it, involved the collection of voluminous *facts* by observation and by trial. The emphasis of the method, as much as the aim of the new philosophy, was on application, experiment, on the practical, as opposed to the purely verbal. The Royal Society of London, which grew out of Boyle's invisible college, adopted the motto *Nullius in verba*—We accept nothing on the mere assertion. In his Utopian novel, *The New Atlantis,* Bacon had envisioned a society governed by inductive philosophers (scientists) met together in "Salomon's House," a kind of combined library, research laboratory, and service club, which would among other things assemble accounts of all trades, crafts, arts, and industries. In imitation of Salomon's House, the young Royal Society sent queries to correspondents all over the world. William Petty, the shipbuilder, wrote histories of the clothing trade, dyeing, and shipping for the Royal Society. Others prepared histories of varnish making, copper engraving, the baking of bread, and the management of forests. From the study of these *facts,* these particular details, general laws were to be drawn. This synthesis of meaningful, simple, summary laws from a mass of detail is the process of induction.

Boyle's discovery of his law—the pressure of a gas is inversely proportional to the volume—illustrates the method. Using mercury to compress the air in a glass tube, Boyle and Hooke made a great many measurements of the different volumes of air compressed by different heights of mercury. They found, for example, that a column of mercury 35 5/16 inches high

* The first printed book in America, *The Bay Psalter,* is dedicated to the Honorable Robert Boyle, President of the Massachusetts Bay Company.

compressed the air in a tube to forty spaces; that $70\frac{11}{16}$ inches of mercury compressed the same air to twenty spaces. Except for $\frac{1}{16}$ of an inch, these observations fit ". . . the hypothesis that supposes the pressures and expansions to be in reciprocal proportion."

Boyle published a table of twenty-five such observations, all of them in the same degree of agreement with his law. For the moment we shall put aside the thought that Boyle undoubtedly

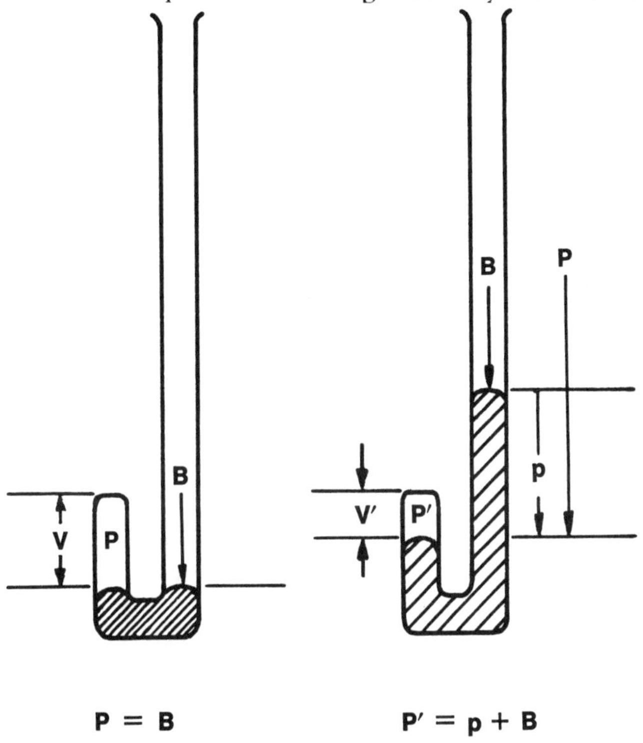

B is barometric pressure, P is the pressure of the air confined within the closed end of the tube.
$P = p + B$.
As P is increased by the addition of mercury in the open end of the tube, V decreases to V′, or P is inversely proportional to V.

Figure 4. Boyle's Law

guessed the law before constructing the tube, and consider only that the inductive process involves the generalization from these or a few particular instances in which pressing on a gas decreases its volume, to a general law which says that these and similar experiments will work in the same way whenever and wherever they may be performed. They will work with other gases than air. They will work at other temperatures than that of Boyle's Oxford laboratory. They will work for other presses than mercury columns and other values than the 70 11⁄16 inches or any of the twenty-five that Boyle published.

But is the law correct? Already in Boyle's statement of it signs of its limitations appear. It is correct only to the degree that the temperature during the experiment remains unchanged. A series of corrections will have to be made before we have accounted for the 1⁄16 of an inch discrepancy. At very high or very low temperatures these correction terms will become large and Boyle's Law will no longer be even approximately correct. The same is true for high pressures or very heavy gases. Nor do we have any guarantee that the physical laws of the universe may not suddenly change completely; merely because all experiments formerly confirmed Boyle's Law, all experiments to be performed in the future may not continue to do so. A physical law predicts the future, and one must be very careful to distinguish between physical laws and crystal balls. Nevertheless, all science is based on the premise that induction is valid. James Clerk Maxwell in 1877 expressed this as the general maxim of physical science: "The difference between one event and another does not depend on the mere difference of the times or the places at which they occur, but only on differences in the nature, configuration, or motion of the bodies concerned."

Isaac Newton in the seventeenth century was more cautious. We cannot be absolutely certain, he wrote, that the time or place of an observation makes no difference:

> In experimental philosophy we are to look upon propositions inferred by general induction from phenomena as accurately or very nearly true, notwithstanding any contrary hypotheses that may be imagined, till such time as other phenomena occur, by

which they may either be made more accurate or liable to exceptions.

It is ironic that the loudest protestations of scientific infallibility emanate from alchemists and gadgeteers, while the Baconian protestations of practicality by Boyle and the Royal Society were to produce an intellectual revolution but little or nothing in the way of material return until long afterwards. Boyle's Law was created by a chain of diffusion, the first of a series of principles often erroneous or imperfectly grasped which nevertheless culminated in the steam engine. The weight of "aire" is a curiously abstract result for a young man and a philosophy which professed devotion to technological and even commercial aims. It led Boyle and his one-time assistant Robert Hooke to a new version of atomism, and to the clear statement two hundred years before Maxwell and in nearly the same terms as Maxwell's of his general maxim of physical science. "Nothing exists," Democritus had written, "but atoms and the void." All phenomena, Hooke wrote, seem merely products of figure, motion, and magnitude.

If the virtuosi of the Royal Society were confirmed Baconians, they were no less influenced by the complementary and partly contradictory philosophy of René Descartes, who if he rejected Aristotelian cosmology went even further than Aristotle in constructing a *deductive* system of the world. Like Socrates in Plato's *Phaedo,* who believed that men were born with the truth which needed only to be brought out by careful consideration, Descartes looked within his own mind for a philosophy and a method of thought which would serve him. "I would close my eyes, stop up my ears, close off all my senses . . ." he wrote, for ". . . I have several times found that these senses are deceiving . . ."

I was then in Germany on the occasion of the wars . . . as I returned from the coronation of the Emperor toward the army, the beginning of winter caught me in a district where not finding any conversation which diverted me, and having besides, happily, no cares or passions to trouble me, I remained all day, closed up in a room with a stove, where I had complete leisure to entertain my thoughts.

As he divested himself one by one of his senses, of all outside stimuli, of all memory, of all that he had learned or previously thought, as he sank deeper and deeper into a state of pure mind —of what, if anything, could he be sure?

"I am certain that I am a being who thinks," he wrote. Descartes' was the method of analysis.

> . . . to conduct my thoughts in order, beginning with the simplest and easiest to know, in order to mount little by little, by degrees to the knowledge of the most complex . . .

Beginning only with the consciousness of self, he believed that it was from this simple (but completely general) assertion that he drew his system, his method, and his science. Of course there were other assumptions in his system, some of which he knew well. He was also well aware that in his reliance upon his own innate (purely mental) knowledge for solving a problem of the physical world (such as the solar system) his solution must be governed by "the great book of the world," that is to say, by experience, experiment, and by the observation and study of physical phenomena.

Boyle's atomism is one example of Cartesian method; Boyle believed that the regular geometric shapes of crystals of salts, which for crystals of the same kind of salt were always similar, reflected a particular configuration of compound particles (corpuscles) of the salt. He therefore suggested that in a reaction between two such crystalline salts in solution, yielding a third salt compounded of the two, the shape of crystals of the third salt should be intermediate between the shapes of the first two.* Boyle's conclusion is a logical deduction from the assumption that, as a mixture of two salts means that such properties as the masses of the salts are added, so also other properties, such as the *shapes,* should also be added and the resultant shapes should be an average of the original two.

Although this conclusion is erroneous as a general law, there are many instances of *isomorphous series* described exactly by Boyle's hypothesis. His observations were here premature. As a Baconian experimentalist, Boyle had made observations of

* Mitscherlich's principle of isomorphism (1819).

particular phenomena—the complex behavior of salts as they precipitate from aqueous solution. They appear as similar polyhedra with plane boundaries. Yet these observations are not of atoms or corpuscles in patterned lattices or otherwise, and Boyle's conclusion from the regularity of the crystal form that this must reflect an internal corpuscular regularity is again logical, mental, rather than experimental. In one sense atoms, corpuscles—the *fine structure* of matter—exist only in our minds because they are beyond the reach of direct observation. Microscopes, no matter how powerful, do not show them. These concepts of the fine structure of matter are inferences, inductive leaps made from the world of phenomena—the world of crystals in geometric forms, or liquids which mix together, or Brownian motion—to the world of ideas, the world of perfect, idealized rules. Logic and mathematics operate on this plane. From Euclid's axioms we draw the first theorems and from these the next. Given that the polyhedral planes of Boyle's crystal salts are caused by regularly disposed arrays of corpuscles, the mathematical averaging of two different forms to produce a third is a Cartesian conclusion.

The scientists of the seventeenth century adopted a thoroughgoing atomism. It was a concomitant of the rising rationalism and mechanism that were to make the next century the Age of Reason. "You know that I held Epicurus strong, and his opinion," Shakespeare's Cassius says before the battle of Philippi. The playwright makes Epicurus stand for a reasoned and ordered world of cause and effect, a world without ghosts, where omens have no meaning.

"We have had of late several comets," John Evelyn, the diarist, wrote after the execution of Viscount Stafford on perjured testimony, "which though I believe appear from natural causes . . . yet I cannot despise them."

Descartes, more than anyone else in the century, had developed the concept of mechanism, of the world as a machine, a clockworks. In the eighteenth century the idea of mechanism grew into deism. The world was a machine, made and set going by God, who had no need to interfere in its works, these proceeding inexorably according to immutable natural law. "I cannot conceive of so beautiful a clock without a maker," Voltaire wrote. It was the skepticism of Epicurus revived and somewhat shaken by two millennia of Christianity.

In men like Franklin, Jefferson, and Madison it influenced the design of the American government—a clockworks with the checks and balances of a clockworks, a government under God as the universe was under God, but operating according to law within the Constitution with church and state strictly separate, as the universe operated according to natural law without the intervention of the Deity.

The atomistic and mechanistic materialism of the seventeenth century was a slow but thorough development dating back to ancient Greece. Atomism and materialism had been a part of the Epicurean philosophy. For Epicurus of Samos (347–271 B.C.) everything, even the mind and soul, was constructed of atoms. Although rationalist and materialist, Epicurus was falsely labeled atheist. His system held gods and spirits but these too were made of atoms—finer atoms than those which made bodies; and because Epicurus was by no means foolish, he had to add a kind of subtle element to his atoms of mind and soul, just as he added elements of heat and breath to his atoms of solid substances.

Later generations did not bother to understand Epicurus. As he had taught that all our knowledge is derived from the senses, this was twisted into an advocacy of unbridled sensualism. The Epicurean philosophy which held that happiness was to be attained through abstinence and moderation was blamed by the later Judaic sects* for the orgiastic excesses of Dionysian cults. The materialism which was a theory of physics became hopelessly confused with a kind of moral materialism of greed and lechery. For the Jews the word for atheist is *epicurean.* In the same way the philosophies of *idealism,* philosophies which subordinate or deny the existence of material bodies and concentrate upon ideas and their relations, are now confused with moral idealism and romantic expectations.

This kind of semantic confusion makes demagoguery pos-

* Christianity, Mohammedanism.

sible and may yet help us to blow up the world. It explains why the seventeenth century began with such hesitation the exploration of atomistic concepts which it was ultimately to embrace. Here too the authority of Aristotle and the schoolmen was directly contradicted, for Aristotle's physics was based on the impossibility of a true vacuum, of nothingness. Matter was continuous and the universe was a plenum. One could divide a stick or a drop of water into parts and these parts into parts again and so on, ad infinitum. Nothing that we may observe or do will bring us down to some last irreducible bit, some atom.

For heterogeneous substances such as flesh and bone Aristotle had even before Epicurus proposed a natural minimum or least subdivision, beyond which a particle could not be subdivided without ceasing to be flesh or bone. These Aristotelian *minima naturalia* were clearly distinguished from *atoms* because they could be cut, as was pointed out later by the Scotists (fourteenth-century scholastic philosophers), and because they did not retain individual identity in larger pieces of matter. The schoolmen who began with Aristotle's method—the logic, the drawing of consistent conclusions from fundamental truths for the establishment of the scholastic Aristotelian system—nevertheless anticipated the Copernican revolution and the seventeenth-century development of mechanics. In the same way their critique of Aristotle led in the seventeenth century to the full revival of the atomic idea.

Averroes, the twelfth-century Moslem Aristotelian commentator, pointed out that the quantity of minima must be related to the size of a body, thus introducing the concept of definite numbers of minima. Using Aristotle's example of a stone being eroded by the steady drip of water, the Averroist Augustine Nifo (1473–1546) wrote:

> Averroes proceeds from the supposition that every increase or decrease consists in the adding or subtracting of a certain number of natural minima.*

Except that the minima were divisible and were themselves different substances, they were for Nifo, who had also attrib-

* Van Melsen, p. 65.

uted local motion* to them, identical with Democritean atoms. Julius Caesar Scaliger (1484–1558) compared hail, rain, and snow, and said that the differences were due to the packing of the minima.

In the *Summary of Physical Science* of Daniel Sennert (1572–1657) of Breslau, the dilemma was resolved by having a higher order of particle to explain chemical combination, a compound particle which he called *prima mista.* In this way the ancients' clear Democritean idea of the original atoms, permanent and immutable through all material change, was by the time of the Renaissance reconciled with the Aristotelian insistence on the change of form. Aristotelian form had become a property of the *prima mista,* or we should say of the molecule.

By the seventeenth century Aristotle was an issue for powerful parties contending for control of the minds of men. The steady erosion of Aristotelian ideas had begun even with their reintroduction in the birth of scholasticism. Modifications and even direct contradictions of Aristotle by such fourteenth-century groups as the followers of William of Occam and the Parisian nominalists were forgotten by the seventeenth-century scholastics resisting the New Philosophy. For the fundamentalists who burned Bruno and condemned Galileo, atomism was in opposition to scholasticism and, like Copernicanism, was another great philosophical heresy of the Renaissance.

Johann Kepler, Imperial Mathematician at the court of Rudolph II, was in 1611 in the midst of the discovery of the laws of planetary motion which bear his name and which establish the Copernican system, when he turned aside briefly to speculate on atomism. In a brief pamphlet, the *New Year's Gift of Hexagonal Snow,* Kepler took up the problem of the geometric regularities in nature: the petals of the flower of the cucumber, the hexagonal cells of the honeycomb, and the geometric forms of crystals. The elegant hexagonal form of the snowflake he considered might be explained by assuming the crystals to be made of tiny round droplets packed in a regular

---

* Motions proper to the individual corpuscles of a body, such as the vibrations of molecules, in modern theory—as opposed to any motion of the whole body, i.e., motion of a car—are called "local" motions.

array. Kepler was too much the neo-Platonist to be satisfied with what he called material necessity. He was led to crystallographic atomism by his passion for geometry and his mystic feeling for reason and order in nature, but he abandoned it because he was too strongly committed to the concept of the world as organism.

We do not meet with Kepler's crystallographic idea again until it appears in the work of Hooke and Boyle half a century later, but for that matter Kepler's astronomical laws were only beginning to be understood at the same time.

Pierre Gassendi (1592–1655), with close connections to both Kepler and Galileo, attempted a revival of the Epicurean philosophy and, as a faithful son of the Church, baptized it, blended its doctrines with Christianity, modifying them where necessary to produce a reconciliation. The universe and all its wonders, material as well as spiritual, are the creation of God. Let God create the atoms and at once materialism is no longer atheistic. Atomism becomes a branch of natural law with Aristotelian logic, alike the gifts and works of God. It was just such a marriage of early Christianity with Greek rationality which had been effected by Aquinas to produce scholasticism in the fourteenth century.

Gassendi revived the Epicurean doctrine that the senses are the source of all knowledge, in opposition to Descartes' concept of innate truth. His atoms were homoiomeric, that is, had properties and parts. One could derive the properties of the atoms from observations of the properties of matter in bulk. Atoms of sticky materials, he suggested, could be like thistles with hooks all over to catch together. Atoms of acids might be spear-shaped or else tiny razors, to cut up the metals which they dissolved. A later generation equipped with the newly discovered microscope was to attribute the sharp taste of vinegar to tiny "eels" biting the tongue. Fibers would be made of string-like atoms, oils of ball-bearing-like atoms. The intellectual absurdity of atoms—the least, irreducible, uncuttable parts of matter—themselves having parts and properties was appreciated by Gassendi as well as by his opponent, Descartes, who himself proposed corkscrew-like atoms for substances which were magnetic. If we only substitute the word molecule for atom, Gassendi's ideas have a curiously modern, popular-science ring to them.

Boyle's corpuscular or atomistic hypothesis was summarized in his *Origine of Formes and Qualities* in 1666. The forms of matter were for the scholastics and the chemists of the seventeenth century the sum of the properties which distinguished a material. Thus gold was characterized by its color, its high density, its luster, its ductility, and its resistance to ordinary acids.

Boyle wrote that there was but one "Catholick or Universal Matter," existing in a kind of fundamental atom of substance which took up space (although very little of it), and was indivisible and absolutely hard (impenetrable). For the Democritean "Nothing exists but atoms and the void," Boyle would have substituted matter and motion. It was the organization and motion of the universal substance into *minima* or *prima naturalis* that gave rise to the apparent diversity of materials. This fine-scale motion he called *local motion,* to distinguish it from the common idea of motion. Boyle conceived the *prima naturalis* as fragments with size and shape, comparable to our ions, whose disposition and ordering gave rise to *texture.* Transmutation of the metals one into another was therefore ". . . no more, than that one parcel of the universal matter, wherein all bodies agree, may have a texture produced in it, like the texture of some other parcel of matter common to them both."

Earlier, Galileo had firmly expressed the basic tenet of mechanism when he distinguished between the primary qualities of physical science, the observables on which reasonable men might be expected to agree—such as length, mass, and time—and the secondary qualities—such as color, taste, warmth —which were in the mind of the observer. A familiar experiment performed in elementary chemistry classes requires three beakers of water, the first icy, the second at room temperature, the third hot. Two students place one hand each in the middle beaker and the other one in the cold or hot beaker. The student with one hand in the hot water reports that the middle beaker holds cold water. The student with one hand in the ice water reports that the middle beaker contains hot water. Such secondary qualities or impressions in the mind, Boyle held, originate in the figure, size, and disposition of the corpuscles.

Because crystals of niter were prismatic, Boyle reasoned that the corpuscles making up the crystal were small prisms whose sharp ends produced the corrosion and the acid taste of the niter in solution. A younger contemporary of Boyle's, Nicolas Lémery (1645–1715), described the neutralization of an acid by suggesting that ". . . limewater softens or diminishes the force of the sublimate because of the particles of lime which it contains, which particles, meeting or hitting the sublimate, break some of the points in which reside the corrosive power."

Besides the *prima naturalis,* Boyle required a higher order of *corpuscle* or *primary cluster,* very roughly comparable to our molecule or radical. These were units having identifiable chemical properties which Boyle could trace through some chemical processes, or changes such as distillation.

Hooke, who had been Boyle's assistant and had become through Boyle's recommendation the preparator-demonstrator for the new Royal Society, had taken over from the young Christopher Wren the society's interest in the newly invented microscope. With this instrument, he wrote, men found at their feet an infinity of new worlds. Each man could be his own Columbus. The world opened up to observation by this new sense (for so Hooke regarded the microscope) he sometimes referred to as "The Invisible Kingdom," although at other times it was the whole world of science which he meant by the phrase.

Under the microscope Hooke found "little living animalcules very prettily a-moving" (protozoa? bacteria?). His careful drawings of the details of the fly's eye, the flea's sting, the louse, the feather, the cells in cork were famous in his time. In every case in which it was possible to observe closely enough, the gross properties of matter could be explained by the finely detailed mechanism revealed under the microscope. Hooke's conclusion of total mechanism, that all the effects of nature, even in those cases in which the mechanism was not yet observable, were produced by tiny machines, was made when he

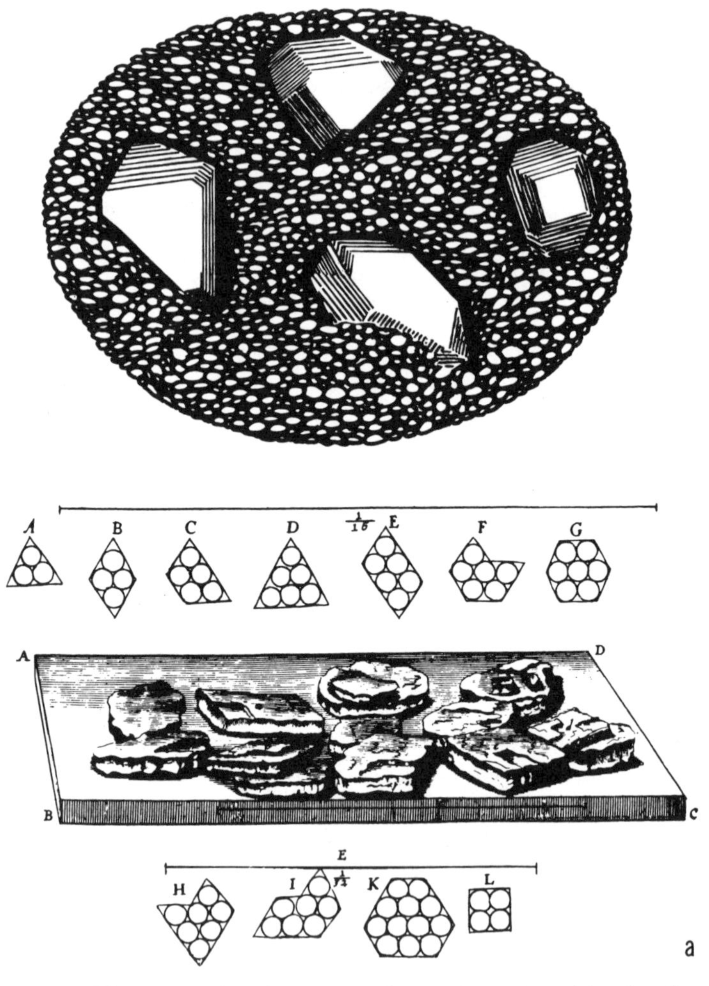

Figure 5. The Plane Faces of Crystals Explained by the Idea of Atoms
a) Robert Hooke, *The Micrographia,* 1666.
b) Christian Huygens, *Treatise on Light,* 1693.

was barely twenty-seven and nothing in the rest of his long scientific life failed to confirm it.

In a series of observations of crystals of quartz and diamond and of salts crystallizing out from solution under the micro-

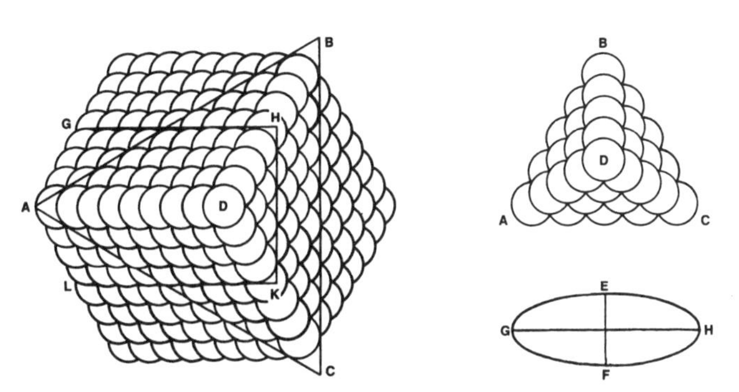

b

scope, Hooke decided that the corpuscles or particles making up these crystals were of neutral shapes—that is, spherical—and that it was their orderly arrangement, rather than any particular shape of their own, which accounted for the regular geometric shapes of the crystals.

Boyle's particles had been homoiomeric—themselves with parts and shapes—but on two levels, the first or *prima naturalis* without chemical properties, the higher corpuscles with chemical properties. Later in the seventeenth century Christian Huygens, the Dutch savant of the Paris Academy of Sciences, deduced from the optical properties of calcite a combination of the two kinds of atomism. Very perfectly crystallized calcite called Iceland spar had just then been discovered to refract (bend) light passing through it in not one but two angles. This birefringence is explained by considering that light passing through the crystal is slowed down more in one direction than another. Huygens, who as a child had known Descartes as a guest in his father's house, adopted Cartesian atomism to explain the birefringence. Descartes' atoms were homoiomeric, like Gassendi's and Boyle's, but Descartes was both atomist and plenist. The holes between his atoms were filled with finer atoms, "splinter matter" ground from the larger ones. There was no void, no empty space possible.

Huygens made his particles of calcite little ellipsoids (flattened spheres) with proportions corresponding to the velocity of light in the crystal. If a bundle of rays is constructed from a point outward with lengths proportional to the velocity of light in the same directions in the crystal, the locus of the ends of the rays forms an ellipsoid. Huygens showed that by packing these ellipsoids together, much as Hooke had packed his spheres, the rhombohedra characteristic of Iceland spar could be built up in any size and the birefringence explained.

The dualism in the atomism of the late seventeenth century is clearly seen in Boyle's attempts to account for his air-pressure experiments. He first postulated homoiomeric particles. The corpuscles of the air might be like the fleece of sheep, that is, each corpuscle would be a tiny elastic spring which could be compressed with its degree of compression proportional to the degree of pressure exerted. Or else the corpuscles might be particles of all sizes and shapes raised by heat into violent motions—an idea of Descartes', but also of Bacon's. In this second case the elasticity or spring of the air is due to the motions of the particles and has nothing to do with their shapes. This is an aniomeric explanation. Both postulates lead by simple mathematics to Boyle's Law.

Hooke's and Boyle's experiments with the air pump, which Hooke elaborated and repeated for the weekly meetings of the Royal Society, had considerable influence on the developments of later science. They established that the air contained a vital spirit necessary to support the life of mice and cats, and necessary to support combustion. By burning gunpowder under water Hooke proved that niter contained its own supply of this vital spirit. It was easy enough to show that insects and birds could not fly in a jar from which the air had been exhausted, and much more clever to demonstrate that air was necessary in order for a bell ringing within the jar to be heard outside. But air, it could be seen at once, was not necessary for the transmission of light, because objects could be seen clearly within a Torricellian vacuum.

With the experiments of Hooke and Boyle the science of chemistry entered a new phase. It had become in the first place an experimental science; in the second place Boyle's Law—which was also found independently by Mariotte (c. 1620–1684)—was the foundation of physical chemistry. It was the first example in chemistry of the kind of mathematically complete relationship which had been the goal of physics since the time of Archimedes. We now know that Galileo did not obtain the foundations of mechanics by dropping stones from the leaning tower of Pisa, and it has even been suggested that he did not perform the experiment of the inclined plane. Galileo was concerned with establishing a consistent and rigorous system of laws of mechanics in mathematical form. Experiment was for the purpose of confirming this system and also, in the event of more than one logically conceivable law, to determine which was applicable to our world. Similarly, more than one relationship between the pressure and volume of a gas is conceivable. The experiments of Boyle, Hooke, and Mariotte were performed to establish a specific mathematical relationship out of an infinite number of possibilities. The inductive method, the emphasis on experience and trial, the recourse to nature as final arbiter which we regard as the foundation of modern science—so much so that most discussions of the history of science begin with the Renaissance—is to be seen principally in Hooke's experiments before the Royal Society, in the parallel experiments of the Paris Academy of Sciences, and in their predecessor societies such as the Accademia dei Lyncei, the Society of the Lynx-eyed, to which both Galileo and Urban VIII, the pope who condemned him, belonged.

The Greeks with their concern for the largest questions had been premature. It was not possible in the early centuries to give meaningful answers to questions of the elements of matter or the least parts into which flesh and bone might be subdivided. The Greek emphasis on rationality and dialectic became in Descartes one half of the scientific method of the Renaissance. The fruitful innovation which was to change the course of human history permanently was the union of the Cartesian deductive and analytic metaphysic with the Baconian inductive method. The remarkable anticipations of the Greeks—particularly in the detailed analysis of the atomistic hypothesis—were nevertheless sterile, because of their relative freedom from the constraints of experience. As far as pure reason could go, the Greeks went. They should be compared with the encyclopedists but not with modern scientists. They aimed for closed systems, for final answers, for facts.

The modern concept of the evolution of ideas, of science as a process involving the continuing dialogue between mind and nature, received its initial impetus in the seventeenth century in the experiments, and to a large extent in the chemical experiments, before the new academies. The motto of the Royal Society, *Nullius in verba,* was not, as it has been described, a revolt against reason, but a revolt against reason isolated from experience. Unlike alchemy, a chemistry could not be built on recipes. Nor could Greek concepts of element and atom prepare gunpowder. It was the union of these—the Baconian gathering of facts, in the light of the Cartesian analysis—which produced the seventeenth-century scientific explosion. The New Philosophy of the Oxford enthusiasts was fruitful because of, and not in spite of, its dependence on the older scholasticism which it was in such haste to destroy.

5

Gases and the Idea of Elements

THE LIGHT WHICH WE HAVE GAINED WAS GIVEN US, NOT TO BE EVER STARING ON, BUT BY IT TO DISCOVER ONWARD THINGS MORE REMOTE FROM OUR KNOWLEDGE.

—MILTON, 1644

When Robert Boyle undertook to bring about the reform of chemistry, by which he meant its conversion to the New Philosophy, he launched his attack on Aristotelianism in *The Sceptical Chymist,* published in 1661. He began by defining as an element ". . . certain primitive and simple, or perfectly unmingled bodies; which not being made of any other bodies, or of one another, are the ingredients of which all those call'd perfectly mixt bodies are immediately compounded, and into which they are ultimately resolved. . . ." The chemists of the early seventeenth century thought in terms of elementary principles, not necessarily material, which entered into the composition of all material bodies and gave them properties such as metallic luster or combustibility. These principles were Boyle's target.

The original Aristotelian universe consisted of a set of concentric spheres with earth at the center, surrounded by the sphere of water which was within the sphere of air which was within the sphere of fire. These made up the material spheres below the sphere of the moon. Beyond were the immaterial, the heavenly spheres with the quintessence, the fifth element of ether, and the celestial spheres culminating in the last sphere, the sphere of Saturn. With the four material elements of earth, air, fire, and water were their associated qualities cold, dry, hot, and wet. In time these four had been expanded to include all conceivable qualities. By the sixteenth century the emphasis of chemistry focused on the qualities rather than upon the elements themselves. Aureolus Philippus Theophrastus Bombast von Hohenheim (c. 1493–1541), who styled himself Paracelsus, was a fifteenth-century physician and mystic concerned with the uses of chemistry in medicine. He founded the school of the iatrochemists, or medical chemists, and altered the classical concept of the elements. He believed that these appeared in bodies in the form of three principles, salt, sulphur, and mercury, a derivation from the old alchemical idea that metals were compounded of the "principles" sulphur and mercury. The common ores of copper, lead, zinc, and mercury are sulfides or compounds of the metal and sulphur, and the experience of the smelters gave rise to the theory.

When Boyle attacked the concept of chemical element, it was these elements and principles to which he objected and for which he substituted his atomistic idea. It should be noted that the iatrochemical concept of element and principle was not that mercury, sulphur, and salt were the elements in any modern sense but that a mercuric *principle,* conferring metallic luster and fluidity, a sulphuric principle or *quality* of combustibility, not NaCl but *salinity* of a material, textural sort, entered into the composition of all bodies.

These concepts of element and principle had also been rejected by Johann Baptista van Helmont (1577–1644), of Brussels, whose work had great influence upon Boyle. Although very little of a mechanist and something of a mystic, van Helmont prepared the way for mechanism and rational chemistry by substituting clear descriptions of his experimentation in place of the hints, confusion, and often deliberate obfuscation of the alchemists. Van Helmont mastered the chemical techniques of the time. He understood the precipitation of one metal from solution by another, weighing the products of the reaction carefully, "because," he said, "nothing is made of nothing," an early expression of the concept of conservation of matter. He experimented with the preparation of acids and salts, preparing and distinguishing sulfuric, nitric, hydrochloric, and acetic acids and aqua regia. He performed extensive experiments with mercury, in which he converted the mercury to a compound and then reduced it to the original weight of mercury.

Van Helmont proposed that only two elements, air and water, made up the whole of the material universe, because fire was immaterial and earth he knew to be made from water. He had proved this, he thought, in the famous experiment of the tree. Here is his account of this experiment as it is given by Partington:

> I took an Earthen Vessel, in which I put 200 pounds of Earth that had been dried in a Furnace, which I moystened with Rainwater, and I implanted therein the Trunk or Stem of a Willow Tree, weighing five pounds; and at length, five years being finished, the Tree sprung from thence, did weigh 169 pounds . . . But I moystened the Earthen Vessel with Rain-water, . . . (alwayes when there was need). . . . At length, I again dried the Earth of the Vessel, and there were found the same 200 pounds, wanting about two ounces. Therefore 164 pounds of Wood, Barks, and Roots, arose out of water onely.

In other experiments, by dissolving vegetable matter in nitric acid van Helmont proved that the process worked in reverse and earth could be converted into water and was therefore not a true element.

We owe the idea of a gas to van Helmont, who tried to burn and heat charcoal in closed vessels. The wasting of wood as it burns he conceived as the loss of its wild spirit, or *spiritus sylvestre*. Wood would not burn in a closed vessel because the spirit could not escape. To this spirit he gave the new name *gas,* or wild gas, apparently from the Greek *chaos,* because he believed it was this spirit which exploded his laboratory apparatus and which gave gunpowder its force. There were many kinds of gases and van Helmont distinguished them carefully as to their origins, whether or not they were poisonous, inflammable, and so on. Of about fifteen gases which van Helmont listed, the majority were simply carbon dioxide.

Van Helmont also considered the experiment of burning a

candle in a glass inverted over water. The rise of water in the glass he attributed to the annihilation by the fire of "something that is less than a body, which fills up the vacuities [holes] in the air." Here was the first setting of the problem of combustion: the *gas sylvestre* and the experiment with the heating of charcoal in a closed vessel led by way of Becher and Stahl to the phlogiston theory. The "something that is less than a body" led by way of Hooke's nitrous air to oxygen and Lavoisier's theory of combustion of elements, and to modern chemistry. Hooke, Boyle, and John Mayow (c. 1641–1679), known as the Oxford chemists, conducted many careful experiments on combustion, and gradually, as a result of these, Hooke became convinced that the air contained a substance, an "air" also contained in niter, which dissolved "sulphureous" bodies (by which he meant inflammable bodies) with the accompaniment of great heat and fire.

The concept of a niter in the air had been prevalent for most of the seventeenth century. John Milton spoke of fire and niter in Satan's fall. It grew from ancient attempts to explain the causes of thunder, lightning, and earthquakes as being the materials—sulphur, saltpeter, or niter—of gunpowder. Gassendi in 1685; Jan Comenius in 1643; Jerome Cardan, the mathematician-physician who was one of the few men with access to the notebooks of Leonardo, in 1550; Vannuccio Biringuccio in his *Pirotechnica* of 1540—all of these and others besides popularized the chemical concept of the union of sulphur and niter to account for geophysical and meteorological violence.

By carefully repeating experiments like van Helmont's heating of charcoal in open and closed vessels, Hooke, Boyle, Mayow, and Richard Lower (c. 1632–1691) were able to establish that their nitro-aerial spirit was necessary for ordinary combustion. With the air pump it could be shown that this spirit was necessary for the respiration of a canary or a mouse. Hooke had a large chamber constructed in which he sat while the air was pumped out from around him, but beyond a slight crackling in his ears, he could detect no ill effects. Luckily for him, seventeenth-century technology was not advanced enough to produce a good vacuum in so large a chamber. In small chambers, however, Mayow found that a gas effervesced from fresh arterial blood but not from stale. In one ghastly experiment, which Hooke could not be induced to repeat for the Royal Society, the upper half of a dog was kept alive for some time by pumping air in and out of the lungs with a bellows. Lower confirmed Hooke's observation that in respiration the blood absorbed a nitro-aerial spirit which turned it red, an observation erroneously rejected by Priestley a century later.

To prove that it was the nitro-aerial spirit in the air and not the air itself that supported combustion, Boyle and Hooke dropped gunpowder on a red-hot iron plate in vacuum and observed it to burn slowly. They also burned gunpowder under water. In order to rule out the possibility that it was admixed air which supported the combustion, Boyle carefully crystallized the niter for the gunpowder under vacuum.

These miscellaneous experiments are not the discovery of oxygen and the modern concepts of combustion and respiration, and merely substituting the word oxygen for nitrous spirit does not suffice to make the ideas of the Oxford chemists into our own. Hooke, who came closest to the modern theory, thought of combustion (and respiration) as a process of solution and the heat evolved as a *heat of solution*. The missing link was the idea of a chemical reaction, but this in turn required the modern concept of elements, which was to evolve over the next century.

Experimentation and the continual examination in rational terms of the explanations of these experiments were the prerequisites for this evolution. Specifically, this generation was characterized by the use of the chemical balance—emphasized by van Helmont and standard procedure by the time Boyle set up his laboratory at Oxford—and the vacuum pump and the techniques for the collection of gases over water and the examination of their properties. By 1700 carbon dioxide, impure carbon monoxide-dioxide mixtures, intestinal gas (impure hydrogen-methane mixture, van Helmont), hydrogen, and nitric oxide (Boyle) had been generated and described as gases. Boyle's experiments with tubes over mercury and water had led to the first of the gas laws, and it was Boyle who first collected a gas over water by inverting a full flask and allowing the rising gas to displace the water.

These studies were continued in a methodical way by Stephen Hales (1677–1761), the vicar of Teddington, who systematically collected and measured gases over water and mercury. Hales did this by heating various substances in a gun barrel and carefully measuring the volume of gas evolved from a given amount of substance. But the concept of gases as distinct chemical species seems to have eluded Hales, who in his thinking if not in his techniques could still be compared with van Helmont, describing as separate gases the carbon dioxide generated in burning charcoal, in fermenting grapes, in acidifying limestone, from spa water, and so on.

Hales noted that his gases differed in solubility, color, inflammability, and smell; but these were simply impurities or taints, he thought, of the fundamental element—air. Hales, who called his book *Vegetable Staticks,* concentrated upon van Helmont's experiment of the tree and demonstrated the transpiration of air through the leaves. Boyle, repeating van Helmont's experiment, had suggested that dust in the air was absorbed by the plant, but Hales demonstrated that the air itself entered the plant and built up its structure. The air which entered the animal and plant vital processes lost its elasticity and became *fixed*. From this condition it could be returned, by distillation in a gun barrel, to its original elastic state, that is, a state in which it obeyed Boyle's Law.

Hales' concept of fixed air was given precise meaning and new significance by Joseph Black (1728–1799). A medical student, Black undertook for his thesis the study of limewater and related substances then believed to cure the dreaded disease of the stone (urinary and gall-bladder calculi). In a remarkable series of experiments he found that heating *magnesia alba,* $MgCO_3$—and later other carbonates—drove off an air which had been fixed in the carbonate and was now made elastic again, obeying Boyle's Law. This fixed air he identified with the gas given off when the carbonates effervesce in acid. The "dead burnt" magnesia, or *magnesia usta,* which would not effervesce in acid, had lost a definite amount of bulk and weight—the weight of the fixed air evolved (although Black did not weigh it).

Black first showed that in a sulfuric acid solution the magnesia usta would regain fixed air from added fixed-alkali (Na_2CO_3) solution and precipitate as the original weight of magnesia alba. He then dissolved weighed quantities of both magnesia usta and magnesia alba in acid, demonstrating that the magnesia alba lost about the same weight in effervescence which had been driven from the magnesia usta by heating. It takes a moment to sum up the experiment and only a few minutes to perform it when the concepts of elements and chemical reaction are clear in our minds, but Black's experiments and the clear thinking that designed them are the first demonstration of a reversible chemical reaction in which all the reactants and all the products are identified and weighed. It took a hundred years to get to Black from Boyle's experiment in which he heated water in a sealed flask for long periods of time and, finding residue in the bottom of the flask, concluded that he had transmuted the element water to earth.

Black's use of Hales' term *fixed air* betrays his concept of his new gas as simply the old element air, now in some way operated on by the inflammable phlogiston principle. Van Helmont's idea of a *spiritus sylvestre* burned away in combustion contained not only the concept of gas, but also the rather obvious idea that combustion involves the loss of a combustible element.

One theory of scientific progress is that there is an obvious common-sense interpretation of every observation or phenomenon and that it is the formulation of these common-sense views that constitutes science. This neo-Cartesian view is implicit in book titles like *Science and Common Sense* and *The Common Sense of Science*. An equally extreme view, which would appeal more to the historian, would be that the obvious common-sense interpretations of phenomena were formulated by Aristotle and that it was for this reason that the scholastic philosophy died so hard. A child believes with Aristotle that heavier bodies must fall faster than light ones and sees that the sun goes around the earth. We are all born Aristotelians and our science is an acquired not an innate taste, much like a taste for caviar or dry wines. We must make an effort of will to visualize the sun as still and the earth rotating; the runaway elevator and the mouse within falling together; the air as made of particles and void. The concept of phlogiston is just such an obvious interpretation of the ordinary combustion of wood and common fuels with which we are all familiar.

Phlogiston (from the Greek *phlogistos,* meaning inflammable) was the base of the theory of combustion of Georg Ernst Stahl (1660–1734). He began his medical career as alchemist, then became professor of medicine and chemistry at Halle, and finally physician to the King of Prussia. Stahl adopted the chemical ideas of J. J. Becher (1635–1682), who would not accept the doctrine of the four elements fire, earth, air, and water, without modification. Intuitively we think of inflammable solids as containing the warmth and light which we associate with fire. Some fuels are richer and hotter than others. Becher had proposed that the element earth, or solid materials, was really threefold, like the principles of Paracelsus—having a mercurial part, a vitreous or fusible earth, and an inflammable or fatty earth (*terra pinguis*). This inflammable earth Stahl renamed phlogiston, a term which had been used in a related context by earlier authors, including van Helmont. Stahl considered that it was "the matter and principle of fire, not fire itself." Combustion was the process of evolution of phlogiston from its combination with other earths. Substances like soot or charcoal were especially rich in phlogiston and could be combined with the residual earths to produce the original combustible matter.

The theory was applied with great ingenuity to the combustion of metals, a subject which had been very thoroughly explored from the earliest beginnings of alchemy. The alchemical metals and alloys, except gold and silver, changed to powdery or crusty dross when heated in open vessels, some, like zinc, burning with flames and smoke. The residue of the combustion of metals was known as the *calx*—and unlike the ash residue from the combustion of wood and fuel, the calx is larger and heavier than the original metal, a fact which must have been observed by the first alchemists. How could it be that the residue of the combustion of a metal, the ash or calx left behind, weighed more than the original metal? The problem was treated quantitatively by Boyle, among others, who thought that the increase in weight was due to the metal's taking up fire particles.

Heating the calx of a metal such as zinc with charcoal, which since it is completely combustible is almost pure phlogiston, yields the original zinc metal. Essentially the idea was that the calx was a kind of elementary vitreous and mercurial earth which could be combined with fatty earth or phlogiston to give a metal. The metal was therefore a compound.

$$\text{Zinc calx} + \text{phlogiston} \rightarrow \text{metallic zinc}$$

Combustion involved the subtraction of phlogiston from the metal, leaving the calx or ash behind as a residue.

$$\text{Metallic zinc} - \text{phlogiston} \rightarrow \text{calx of zinc}$$

Charcoal consisting of ash plus phlogiston could not be recombined after combustion, but that was simply a detail. A more serious objection was that the calx, though not the ash, weighed more without phlogiston than the metal with phlogiston. Stahl held that phlogiston reduced the weight of substances. Others held that phlogiston was lighter than air, like hydrogen. Cavendish in 1766 discovered hydrogen and suggested that it was phlogiston.* Still others maintained that phlogiston had negative weight, or levity, rather than gravity.

Stahl's phlogiston theory was so simple, such a beautifully clear exposition of common and well-known experiments, so consistent and so successful in leading to further experiment and discovery, that it seems almost cruel to suggest that it retarded scientific chemistry for nearly a century. There is no sadder illustration of the fallacy that science consists simply of the observation of "facts" and their assembly into meaningful order, or that the "truth" which the scientist seeks is the

* Hydrogen, like charcoal, will reduce an ore or calx to metal.

same uncomplicated "truth" we sought as children. For we are not children and our race is no longer in its infancy.

The trap of phlogiston, into which the chemists of the Age of Reason almost without exception fell—the principal nonphlogistonite was Hermann Boerhaave (1668–1738)—had been easily avoided by the Oxford chemists much earlier, for with all their Baconian protestations they were well armored by their atomistic preconceptions.

"I have imposed upon myself, as a law, . . . never to form any conclusion which is not an immediate consequence necessarily flowing from observation and experiment," wrote Lavoisier, the man who upset the doctrine of phlogiston. Yet it was just this devotion to the simple, direct phlogiston explanation of combustion and the calcination of metals which betrayed most of the chemists of the eighteenth century. Lavoisier, driven by the urge to distinguish himself, ranged himself against the phlogiston theory and soon found enough illogicalities and inconsistencies to attack. But the principal evidences which he used in his attacks were those gathered by the phlogiston chemists, beginning with Joseph Black.

After his first researches on carbonates Black extended some of Hales' observations and showed that "the change produced on wholesome air by breathing it, consisted chiefly, if not solely, in the conversion of part of it into fixed air. For I found that, by blowing through a pipe into lime water, the lime was precipitated. . . ." The same test showed that another of van Helmont's gases, that produced in the brewing of beer, was identical with fixed air and, as van Helmont had conjectured, with the gas produced in burning charcoal and with the poisonous vapors of the Grotto del Cane in Italy (actually containing admixed carbon monoxide).

The real significance of Black's work, however, was not in the discovery of carbon dioxide, which had been known to every chemist since van Helmont, nor in the determination of its chemical and physiological properties, which also antedate Black in part, nor in the determination of the physical properties of the gas, which waited for his disciple, Henry Cavendish, nor in his concept of a gas—since he continued to regard his fixed air as a part of the ordinary element air which had been converted to a distinct species, perhaps by the operation of phlogiston. Black's contribution—"a new, and perhaps boundless field seemed to open before me," he wrote—did indeed found English pneumatic chemistry. It made possible the discoveries of about a dozen gases by Rutherford (1749–1819), Scheele, Cavendish, and Priestley, which in turn led to the founding of modern chemistry by Lavoisier. It was the proof by reasoned quantitative experiment that a gas was a state of matter which entered into chemical combination with other matter, completely transforming all properties except mass, and which by dissolution could be obtained again entirely in its original form.

" . . . it seems probable to me," Newton wrote not long before his death in 1727,

> that God in the Beginning form'd Matter in solid, massy, hard, impenetrable, moveable Particles, of such Sizes and Figures, and with such other Properties, and in such Proportion to Space, as most conduced to the End for which he form'd them; and that these primitive Particles being Solids, are incomparably harder than any porous Bodies compounded of them; even so very hard as never to wear or break in pieces; no ordinary Power being able to divide what God himself made one in the first Creation. . . . And therefore, that Nature may be lasting, the Changes of corporeal Things are to be placed only in the various Separations and new Associations and Motions of these permanent Particles . . .

But in the complexity of a material world in which 500,000 formulae could not express all the compositions of nature, what seemed a probability to Newton could not even in part be demonstrated until Black began the work.

Henry Cavendish (1731–1810), who was later to demonstrate his experimental virtuosity by measuring the force of gravity, seized upon the concept, which Black had noted almost as an aside, of distinct species of elastic fluids-gases. He used the term *factitious air,* taken from Boyle, for "any kind of air which is contained in other bodies in an unelastic state and is produced from thence by art," and proceeded to measure exactly the properties of these airs. It was Cavendish, not Black, who first weighed the amount of fixed air driven off marble by the action of acid and, finding that the fixed air was soluble in water, introduced the practice of collecting gases over mercury. This had been used previously by Hales, but it was scarcely noticed and not exploited systematically until Priestley later devised his pneumatic trough. Boyle had already collected an inflammable air (hydrogen) given off by the action of dilute sulfuric acid on iron. Now Cavendish collected this air and measured its density (fourteen times lighter than common air) and its other properties, finding that it formed an explosive mixture with air: "I know of only three metallic substances, namely zinc, iron, and tin, that generate inflammable air by solution in acids; and those only by solution in the diluted vitriolic acid [H_2SO_4], or spirit of salt [HCl]."

Just as the action of these acids on the magnesia alba or limestone drove fixed air from its combination with magnesia usta or quicklime, Cavendish interpreted his experiments to mean that "when . . . the above-mentioned metallic substances are dissolved in spirit of salt, or the diluted vitriolic acid, their phlogiston flies off . . . and forms the inflammable air."

These experiments not only isolated and identified a second species of gas—hydrogen—but confirmed beautifully and clearly the phlogiston concept of a metal as a compound of a calx and phlogiston. That Cavendish, an experimentalist so neat as to be able to measure in the laboratory the gravitational attraction of two lead spheres, should have missed the fateful implications of the increase in the weight of the calx of a metal is a testimonial to the power of the formalist solution of the problem.

The historian Herbert Butterfield once compared the science of chemistry to a board game. Let us suppose for one brief moment that metals, hydrogen, oxygen, acids, and so on do not exist in reality but that we are engaged in the solution of an intellectual puzzle. We have a board on which we may introduce counters, red, yellow, blue, or knights, bishops, and pawns, or even counters which *do things,* such as dice or dials. From above the board we observe the erratic behavior of the pieces. A pawn reaches the end row, disappears, and a queen takes its place. A red counter moves diagonally while dice turn up the number six. We introduce a blue counter and it leaps rapidly about the board, wiping half the pieces into oblivion. But much of the board remains cluttered. A dial still turns up numbers; the dice continue to roll; a card turns up reading "Go back to O." Our blue counter clearing half the board is the concept of phlogiston. As the game progresses, the blue counter again and again triumphs over half the pieces which are brought into the game. Against the others it can do nothing. When we introduce the red counter—Lavoisier's concept of oxidation—nearly all the pieces will fall into place. We shall understand the function of the dial reading and the cards as they turn up. But as we look about the board we see that it is indeed very large and may in fact be infinitely large. Here are areas with pins and balls moving among them and lights that flash and a multitude of counters immune to the operations of the red counter of Lavoisier.

The supposition is of course false. Metals, hydrogen, oxygen, atoms exist. Phlogiston does not exist. We know that a metal exists because we can see and touch it—that is because it reflects light which all of us except the blind are equipped to detect (and even the blind may be equipped with a suitable photoelectric cell rigged to a bell, so that they may *hear* the light reflected from the metal). Hydrogen exists because, although it does not reflect light to our eyes, a few simple operations in a laboratory will produce an explosion, noise, light—all of which we interpret in a complicated way as "proof" of the generation of hydrogen. But then phlogiston exists, since the application of logic to the experiment yields also a "proof" of phlogiston.

Now nothing is to be gained at this point by a revival of the theory of Becher and Stahl, just as nothing is to be gained by a return to the universe of Aristotle with the planets and the sun going around the earth. But we have known since the time of Ptolemy that a mathematically consistent picture of such a universe, agreeing perfectly with observation, could be constructed. And in fact, according to the general theory of relativity, there can be no privileged point from which to observe the universe, and the choice of geocentric or heliocentric universe is simply one of convenience and not one of truth or falsehood. The ship's navigator who speaks of the stars rising and setting is using the Aristotelian concept. With his attention restricted to one area of the board, this counter works best. The classical physicist, who also has laws of motion to consider, must use a Copernican counter. Only in a few exceptional experiments—experiments involving areas of the board far removed from the ordinary—does he find a failure of the laws of motion and the necessity to fall back on a relativistic counter. Similarly, it would in theory be possible to construct a phlogiston chemistry. It would immediately require new and strange laws of physics—substances with levity, for example, repelling rather than attracting matter. But the history of recent physics has shown that mathematical laws can be constructed to meet the requirements of experiment. Since our science from the death of phlogiston onward has been constructed on different lines, the task of devising a consistent phlogiston chemistry would require an enormous and totally impractical revision in every field. Phlogiston was not a very *good* counter. It was the most obvious—it lay on top of the pile.

It was Lavoisier who saw the moves which would swiftly clear the board. "I hardly know of any person, except my friends of the Lunar Society at Birmingham, who adhere to the doctrine of phlogiston; and what may now be the case with *them,* in this age of revolutions, philosophical as well as civil, I will not at this distance answer for," Priestley wrote sadly from America. Phlogiston was a doctrine, an *idea* ". . . at one time thought to have been the greatest discovery ever made in the science."

The idea was not *wrong;* one might as well say that a car or a bridge is wrong. It is a construct, and the ideas implied in the terms hydrogen, oxygen, atom, acid are similarly constructs, abstract ideas which try to provide intellectual comprehension of the complexity of our experience.

6

The Triumph of the Antiphlogistians

THE ONE, CIRCUMSPECT PHYSICIST, EXAMINES ONLY OBJECTS IN THE DOMAIN OF EXPERIENCE, BRINGS TO HIS PROCEDURES ONLY A TIMID AND RIGOROUS LOGIC; PERMITS HIMSELF NEITHER SYSTEMS NOR PREJUDICES; SEARCHES ONLY FOR TRUTH WHATEVER IT MAY BE, AND ALMOST ALWAYS DISCOVERS AND ESTABLISHES IT IN THE MOST SOLID AND BRILLIANT MANNER. THE OTHER, RASH THEOLOGIAN, APPROACHES WITH AUDACITY THE MOST MYSTERIOUS QUESTIONS, SCORNS THE BELIEF OF CENTURIES. . . .

CUVIER IN THE *Éloge* OF JOSEPH PRIESTLEY

Joseph Priestley (1733–1804) is an excellent example of the eighteenth-century man of reason, once so strongly admired but currently out of fashion. Priestley's was a tender mind, Lavoisier's a tough one. Priestley wrote no less than one hundred and eight works, prolix, rambling, imaginative, and speculative. Raised in the radical religious ferment of the English clothmakers of the Midlands, Priestley was proficient in Latin, Greek, and Hebrew, as well as modern languages. His Unitarian convictions cost him his successive posts as minister and schoolmaster until, settling in Birmingham, he found himself in the congenial company of Boulton and Watt, makers of steam engines, Josiah Wedgwood, the founder of the great English potteries, and Erasmus Darwin, grandfather of Charles* and author of a theory of evolution written in verse.

* Wedgwood was Charles Darwin's maternal grandfather.

The scientific club—the Lunar Society—in which they met can almost be compared to the early Royal Society in its fertility and influence. But with the outbreak of the French Revolution the republican sympathies of these eighteenth-century intellectuals brought down the wrath of the mob and Priestley was driven reluctantly to exile in America.

In 1791, on the anniversary of the fall of the Bastille, while the authorities stood by, the mob burned his house and laboratory. Warned by friends, Priestley and his family and servants had escaped. The new engine-works of Boulton and Watt were also threatened. "We on our part, finding there was no likelihood of any other protection, applied to our workmen, . . . procured some arms, and had their promise of defending us and themselves against all invaders," James Watt wrote of the night which saw the razing of Priestley's home.

Priestley had perfected chemical apparatus for the collection and study of the properties of gases over water and mercury. With his pneumatic trough he could study the effect of gases on mice and plants, test for inflammability, for solubility, and measure volume and weight. His test for the "goodness" of the new "dephlogisticated air" (oxygen) which he had just discovered was the length of time a mouse could live in it.

"In this air," he wrote, "my mouse lived a full half hour; and though it was taken out seemingly dead, it appeared to have been only exceedingly chilled; for, upon being held to the fire, it presently revived, . . . it remained perfectly at its ease another full half hour, when I took it out quite vigorous."

In 1772 Priestley published his discoveries of nitrous air and nitrous vapor (nitric and nitrous oxides) obtained from the action of nitric acid on copper. He reported a marine acid air (HCl from the action of hydrochloric acid on salt) and carbon monoxide. Nitric oxide had been prepared by van Helmont and collected by Boyle, among others. Cavendish had prepared the marine acid air which had immediately gone into solution in water, but Priestley collected it for analysis by using mercury in his trough. Later he obtained vitriolic acid air (SO_2) and fluor acid air (silicon fluoride) by similar procedures. This last gas was also obtained by Scheele. From spirit of hartshorne (ammonia solution) he obtained ammonia gas (NH_3) and, in testing the inflammability of this by passing electric sparks through its container, decomposed it into hydrogen, which he identified as Cavendish's inflammable air and "phlogisticated" air (nitrogen). He identified the phlogisticated air with the residue of the air left after the calcination of a metal. There was almost a race at this time among the growing number of chemists to produce and study the seemingly inexhaustible variety of "factitious airs."

Carl Wilhelm Scheele (1742–1786), a Swedish apothecary devoted to experimentation, concentrated upon the contraction undergone by a volume of air in contact with phlogiston-rich water mixtures, such as iron filings in water. This led him to describe ordinary air as a mixture of ". . . two fluids, differing from each other, the one of which does not manifest in the least the property of attracting phlogiston. Whilst the other, which composes between the third and fourth part of the whole mass of the air, is peculiarly disposed to such attraction." These were nitrogen and oxygen, called respectively *foul* and *fire* air by Scheele.

Before the news of this discovery of oxygen reached England, however, Priestley in 1774 had obtained a twelve-inch burning glass. He focused the sun's heat in beakers of various chemicals supplied to him by "Mr. Warltire, a good chymist" in the attempt to find new airs. The red calx of mercury yielded an air which by the intensity with which it supported the combustion of a candle flame, and by the length of time it sustained his poor mouse, convinced him that it was exceptionally devoid of phlogiston—therefore dephlogisticated. ". . . This pure air may become a fashionable article in luxury . . ." he wrote. "Hitherto only two mice and myself have had the privilege of breathing it." It was the discovery of dephlogisticated air that precipitated the revolution in chemistry and the birth of our modern concepts—a revolution resisted to the end by Priestley.

Antoine-Laurent Lavoisier (1743–1794) was the architect of this remarkable revolution and the victim of the political revolution with which it approximately coincided. If physics had its martyr in Bruno, chemistry was to have its martyr in Lavoisier. A monarchist mob hounded Priestley into exile, a republican tribunal condemned Lavoisier—legend has it with the words, "The republic has no need of scientists." Borda and Haüy, members with Lavoisier of the Commission of Weights and Measures of 1790, which founded the metric system, risked their own lives to appeal for Lavoisier's life, but the old Academy of Science had treated contemptuously the youthful scientific work of the republican hero Marat. Lavoisier's disciple and colleague Antoine Fourcroy (1755–1809), now in Marat's Assembly seat, called for a political purge of the Academy.

Two years later the Jacobins had fallen and Fourcroy himself delivered the sad memorial speech for Lavoisier. "Carry yourself back to that frightful time," said Fourcroy, who had prospered mightily during the period in question, ". . . when the least word, the slightest mark of solicitude for the unfortunate beings who were preceding you along the road to death, were crimes and conspiracies." Just so in Germany today we find no former Nazis, in Russia no Stalinists. The scientists who rose to rank and position while their colleagues were driven to exile or prison or murderous death accepted their positions and worked for the regime only to prevent worse things from happening. In reality and in their hearts they were always against the regime. . . .

Lavoisier was one of those young men who is described as "promising." A youth of charm and good family, at twenty-one he accompanied J. E. Guettard (1715–1786) in the field studies for one of the earliest geological maps. At twenty-three he received a medal from the Paris Academy for an essay written with full attention to literary style as well as content, and in the same year was elected to the Academy. It would appear that from the beginning in those revolutionary times, Lavoisier aimed at nothing less than a "revolution in physics and chemistry. I have felt bound to look upon all that has been done before me merely as suggestive," he wrote.

> I have proposed to repeat it all with new safeguards in order . . . to form a theory. The results of the other authors whom I have named, considered from this point of view, appeared to me like

separate pieces of a great chain; these authors have joined only some links of the chain.

The reader will recall Descartes.

Experimenting with factitious airs driven from matter, the young Lavoisier ("I was eager for glory") stumbled on the increase of weight of sulphur and phosphorus* in combustion. Although it was then commonly known that metals increased their weight in calcination, it had been generally assumed that ordinary inflammable substances *lost* weight with the loss of phlogiston. The anomaly suddenly struck Lavoisier with such force that he confided his results in a sealed note to the Secretary of the Academy (November 1, 1772), to establish his priority to what he immediately understood to be the discovery of the century.

> . . . this increase of weight arises from a prodigious quantity of air that is fixed† during the combustion and combines with the vapours. . . . I am persuaded that the increase in weight of metallic calxes is due to the same cause. . . . I have carried out the reduction of [lead oxide] in closed vessels, with the apparatus of Hales, and I observed that just as the calx turned into metal, a large quantity of air was liberated. . . . this discovery appearing to me one of the most interesting of those that have been made since the time of Stahl, and since it is difficult to prevent something from slipping out in conversation with friends which might put them on the track of the truth, I have thought it right to make this deposition . . . against the time when I shall publish my experiments.

Had Priestley been as cautious in confiding his work to sealed letters, Lavoisier would certainly have suffered great delay, but the author of one hundred and eight publications described his dephlogisticated-air experiments of 1774 with great enthusiasm at Lavoisier's dinner table. At once Lavoisier realized that Priestley's air was the very air which he had described as fixed in the combustion process. Not realizing

* Observed by Marggraf in 1740.

† Expressly stated by Jean Rey in 1630 for tin and lead, by Hales in 1727 for sulphur and phosphorus.

until later (probably after reports of Scheele's work reached Paris) that another gas (nitrogen) made up four-fifths of the bulk of the air, Lavoisier at first thought that the dephlogisticated air was common air made "more respirable, more combustible, and consequently that it was more pure than even the air in which we live." He had still not attained to the concept of air as a mixture of distinct chemical species. He thought of dephlogisticated or pure air as a compound of "the matter of fire or light" with a base, saying that the phlogiston theory explained combustion "in a very happy manner" but that it placed the matter of fire (phlogiston) in the combustible, while he placed it in the air.

These researches were brought to a triumphant conclusion, not only replacing the phlogiston theory but finally overthrowing the old doctrine of the four elements, when Cavendish discovered that water was a compound of hydrogen and oxygen. Very early, Lavoisier, always precise and careful in his experimentation, had repeated the alchemical transmutation of water into earth which had convinced Boyle of the possibility of transmutation. Water was boiled indefinitely in a sealed glass vessel (pelican), which was then opened. Upon evaporation of the water some sediment was left behind. Lavoisier was the first to compare the total weight of the sediment with the weight of the pelican before and after the experiment, thus demonstrating that the sediment had been dissolved from the glass. This was confirmed when Scheele learned to analyze the sediment. Lavoisier had repeated the experiment with the conviction that matter (mass) must be neither created nor destroyed and that therefore "an equal quantity of matter exists before and after the operation." This principle, which is the essence of materialism, had been sharpened by van Helmont and Boyle, but especially Black, from whom Lavoisier took it.

Heating tin in an open vessel, Boyle found an increase of weight which he attributed to fire particles passing through the glass into the tin. At Boyle's stage of philosophical materialism it had not seemed unreasonable to attribute weight to heat and light, although Bacon had described heat as local motion and Descartes and Hooke had proposed the modern idea of heat as a ceaseless motion of particles. But Lavoisier sealed his tin in a glass vessel and showed that there was no change in the weight of the whole apparatus until, when the seal was broken, air rushed into the vessel to take the place of the air consumed in calcination, with an appropriate increase in weight. "It is no less true in physics [chemistry] than it is in geometry that the whole is equal to the sum of its parts," Lavoisier wrote. It was the law of conservation of mass, now called the principle of Lavoisier, for it is in the work of Lavoisier that we first see it as an axiom of chemistry.

At every turn in the account of the rise of the new chemistry we see the fine hand of Priestley. His were the experiments, his were the discoveries, his the introductory ideas—"a tissue of experiments unconnected by the least thread of reason," Lavoisier said. Yet it was the prolix Dr. Priestley, the Socinian parson, his head a jumble of phlogiston and belief in the goodness of man, whose laboratory provided those equalities and those gases on which Lavoisier built his theories. Priestley had been originally turned to science after his meeting with Benjamin Franklin, just such another tender-minded freethinker, and his first researches were in the field of electricity. It was natural for him to try to see what factitious airs could be released through electricity. With Warltire, Priestley exploded air and inflammable air (hydrogen) together in closed vessels, noticing the deposition of dew, but he was misled by Warltire into thinking that water must be a constituent of air.

He communicated these experiments to James Watt, to Cavendish, and to Lavoisier. Cavendish, of whom it was said that "his theory of the universe [was] that it consisted *solely* of a multitude of objects which could be weighed, numbered, and measured," repeated the experiments with great care. By 1783 he had proved that the two airs combined to make pure water. To appreciate the skill of these experimenters it should be recalled that under these conditions small amounts of nitric acid are formed from traces of nitrogen in the gases—a separate chemical problem which Cavendish solved completely.

Watt also obtained water from the union of hydrogen and

oxygen,* but he explained it in the light of the phlogiston concept. The inflammable air which the English pneumatic chemists had previously believed to be pure phlogiston they now took to be a compound of water plus phlogiston, while dephlogisticated air (oxygen) they now took to be water lacking phlogiston. Combining the two yielded pure water. One might even assign suitable weights to the phlogiston and balance the experimental equation. But although the phlogiston doctrine could again be patched by slight modification here and there to fit the new experiments, the nomenclature began to creak. Dephlogisticated air would have to be changed to dephlogisticated water. It was in the revision of nomenclature that the English really lost. "Whether we adopt the new system or not," Priestley protested, "we are under the necessity of learning the new language, if we would understand some of the most valuable of modern publications."

Learning of Cavendish's experiments in 1783, Lavoisier and Laplace the mathematician burned inflammable air (hydrogen) in dephlogisticated air (oxygen) before a group of academicians:

> . . . as it was not possible to assure ourselves of the exact quantity of the two airs with which we had thus accomplished the combustion; but as it is no less true in physics than it is in geometry that the whole is equal to the sum of its parts . . . we believe ourselves correct in concluding that the weight of the water was equal to that of the two airs which served to form it.

Lavoisier in his urge to systematize, to theorize, to make of the jumble of chemistry a coherent unified whole like Newton's mechanics—indeed, it was his ambition to become the Newton of chemistry—continually leaped to conclusions from these experiments of others—as he had when he first thought that Priestley's dephlogisticated air was common air. He built whole systems on these conjectures, and one of these was his conviction that pure air—dephlogisticated air—was the *acidifying principle,* or in Greek derivative, *principe oxigene,* later *gaz oxygene.* He meant by this a substance which would combine

* Also independently discovered by the mathematician Monge.

with other substances to form an acid, an idea which he obtained from a systematic study of acids and which only his study of HCl stubbornly failed to confirm. When he heard of Cavendish's experiment Lavoisier shifted the emphasis in his theory from the acidifying effect of oxygen to the relatively simple concept of combustion as a chemical combination with oxygen.

The use of the new terminology blinds us to the distinctions between our modern concepts and those of Lavoisier. He still regarded light and heat—*calorique,* "caloric"—as material if imponderable or weightless substances and the "matter of fire"; and oxygen as a chemical compound of these elements with a ponderable base.

In a significant sense, phlogiston was not overthrown. It was renamed caloric and relocated in the air instead of the metal. Lavoisier was not the champion of truth confronted with error but rather a man of brilliance driven by pride and the thirst for glory to a dangerous prominence. If we read it now, his new chemistry is full of the same faults with which he reproved the phlogistonites. ". . . Now this principle [phlogiston] has weight and again it is weightless . . . now it is free fire and again it is combined with the element earth; Now it penetrates right through the pores of vessels, and again it finds bodies impenetrable," he wrote in his *Traité de Chimie.* Yet Lavoisier saw no inconsistency in defining free- and combined-caloric as loosely as the phlogistonites had defined their fluid. Here he defines caloric:

> It is difficult to comprehend these phenomena, without admitting them as the effects of a real and material substance, or very subtile fluid, which insinuating itself between the particles of bodies, separates them from each other; and, even allowing the existence of this fluid to be hypothetical, we shall see in the sequel that it explains the phenomena of nature in a very satisfactory manner.

It was more than naïveté that prevented Priestley from joining the followers of Lavoisier in decrying the phlogiston concept. Lavoisier, the apostle of precision and laboratory measurement, explained others' results. He had yet to find one new chemical species, discover one new reaction. His theory of acids formed by oxygen had from the beginning one significant exception, HCl. It broke down with the first new experiment (Cavendish's combustion of hydrogen with oxygen). His caloric was a modified and renamed phlogiston. His theory did not explain a second "heavy inflammable air" (CO) which Priestley had discovered.

With hindsight, it is easy to see that caloric was not necessary to Lavoisier's chemistry and that the advances of physics would soon leave no choice between the two theories of combustion. But it was in Lavoisier's lifetime, even before the Cavendish experiment, that the ascendancy of this chemistry was assured by the development of a remarkable terminology. This terminology, which we now use, contains within itself the modern ideas of element and reaction, as well as Lavoisier's concept of combustion, and divides us most effectively from all earlier chemistry. Butter of antimony, sugar of lead, liver of sulphur, Glauber's salt, regulus, sal alembroth, powder of algaroth—yet even this eighteenth-century terminology represented an advance over the deliberate obfuscation and occultism of the alchemists. Lavoisier, Guyton de Morveau (1737–1816), C. L. Berthollet (1748–1822), and Fourcroy were appointed by the Academy to revise chemical nomenclature completely, and they reported back in 1787. They understood that the lexicographer was mightier than the experimenter.

> While engaged in . . . reforming the nomenclature . . . I perceived, better than I had ever done before, the justice of the following maxims of the Abbé de Condillac, in his *Logic,* and some other of his works.
>
> "We think only through the medium of words,—Languages are true analytic methods.—Algebra, which is adapted to its purpose in every species of expression, in the most simple, most exact, and best manner possible, is at the same time a language and an analytic method.—The art of reasoning is nothing more than a language well arranged."

The analogy is to a bridge, a construction, rather than to a map or a description. Chemistry from there on took the form and direction it did because it had been pointed in this direction by Lavoisier's nomenclature, and not because this was "the way it was." Guyton de Morveau supplied Lavoisier with the system of compound words. An acid (according to Lavoisier's mistaken theory) consisting of a base with oxygen would be known by the name of the base, as for example, *sulphuric* acid (replacing vitriolic acid). If the same base would form another acid by combining with less oxygen, the termination would indicate the lesser state of saturation, as sulphur*ous* acid. Salts formed from sulphuric acid would be *sulfates;* from sulphurous acid, *sulfites.* "Sulphide will denote all the compounds of sulphur not carried to the state of an acid."

The committee published a dictionary listing seven hundred chemicals with their old names translated into new ones, expressing immediately the constituents of a compound body, as *ammonium molybdate.** The word *gas* Lavoisier adopted from Macquer's *Dictionnaire* of 1766. The four elements, the three principles, the airs, phlogiston, are no longer with us.

> All that can be said upon the nature and number of elements is, in my opinion, confined to discussions entirely of a metaphysical nature. The subject only furnishes us with indefinite problems, which may be solved in a thousand different ways, not one of which, in all probability, is consistent with nature . . . if we apply the term *elements* or *principles of bodies,* to express our idea of the last point which analysis is capable of reaching, we must admit, as elements, all the substances into which we are capable, by any means, to reduce bodies by decomposition.

With Lavoisier's *Traité de Chimie* in 1789, matter was classified into five simple substances "which may be considered as the elements of bodies." These were light, caloric, oxygen, azote (nitrogen), and hydrogen. There were also six oxidable and acidifiable simple substances: sulphur, phosphorus, charcoal, and the muriatic,† fluoric, and boracic radicals; seventeen

* I.e., the ammonium salt of the fully saturated acid of the base molybdenum.

† Muriatic acid is hydrochloric acid. Lavoisier's erroneous theory of acids led him to ignore Scheele's discovery of chlorine gas. Davy in 1810 finally proved the absence of oxygen in HCl, and gave Scheele's gas the name chlorine.

metals, arranged alphabetically according to their French names; and five "salifiable, simple earthy substances," lime, magnesia, barytes, argill (clay), and silex.

Like all true revolutionaries, Lavoisier was neither so radical nor so in opposition to his past as he pretended. It is apparent that his whole concept of combustion and reaction derived from the previous theory, the phlogiston theory, and was conceived in its spirit and its method, representing at the same time a major simplification of that theory. Similarly, the phlogistonites, and Priestley especially, provided by their researches and continuing critique the possibility of the new system. From the perspective of two centuries the progress of material science shows not so much an explosion in the time of Lavoisier, as a kind of flow much like that of a stream which has been placid enough but now emerges onto new terrain and, nourished by the influx of other streams, gradually quickens its pace.

We who are in the position of being swept away by the torrent which science has become too frequently seek to account for our predicament by the easy explanations of individual genius. Certainly the pace of science in Lavoisier's time was quicker than it had been before, and after Lavoisier it was to be quicker still. But the new system which Priestley deplored was the legitimate heir as much as the conqueror of the earlier chemistry, and the old passed peaceably into the new, the continuity between the two masked by the personalities of a society that was about to produce a Napoleon. Nor may we say that the new system was, except in embryo, the modern system; but rather that, like the phlogiston system before it, it provided those basic ideas which were ancestral to our present concepts. No one expressed these better than Lavoisier, who used the tool of his incisive mind to carve a logical system from the almost formless mass of data and results of his time. It is to Lavoisier that we owe the operational definition of an element as the limit which may be reached by analysis.* His was the driving force behind the development of a rational nomenclature and

* Cf. page 77, Boyle's ideal definition as the perfectly simple limit of the resolution of matter out of which all other forms of matter may be compounded.

the organized system of ideas of his *Traité,* after which, whatever emendations, corrections, and outright contradictions were still to be added, the science of matter had been transformed irreversibly and stamped for all future time with the impress of his mind.

7

The Laws of Classical Chemistry

SINCE, THEN, ALL OTHER THINGS SEEMED IN THEIR WHOLE NATURE TO BE MODELLED AFTER NUMBERS, AND NUMBERS SEEMED TO BE THE FIRST THINGS IN NATURE AS A WHOLE, THEY [THE PYTHAGOREANS] SUPPOSED THAT THE ELEMENTS OF NUMBERS ARE THE ELEMENTS OF ALL THINGS, AND THAT THE WHOLE HEAVEN IS A MUSICAL SCALE OR NUMBER.

—ARISTOTLE

As later societies were to look back after the breakdown of the Dark Ages and regard the world of Greece and Rome with romantic nostalgia, looking to its standards with envy of that self-confidence born of accomplishment, so the modern world of science looks back on the nineteenth century and its confidence in mechanism and materialism. This is the classical world for the community of science. The laws of mechanics were the canons for a society which saw all phenomena, all without exception, as ultimately reducible to movement, shape, and number—Robert Hooke's motion, figure, and magnitude. The eighteenth century had expressed this fundamental assumption in the form of Newton's laws of motion—the simple assertions that a body in motion remained in motion and that a force was required to charge that motion; and that actions were accompanied by equal reactions. Phenomena such as the precise form of the moon's orbit could be expressed completely in mathematical equations relating force, time, and distance. These equations in turn could be derived directly from Newton's laws. If the eighteenth century was the Hellenic period of science, the nineteenth was the Hellenistic. The simplicities of Newton's laws and his plane geometry were supplanted by a hundred years of the development of his calculus.

Chemistry in the early nineteenth century was in the process of replacing its earlier alchemical traditions of trial and error, of purest empiricism, with rigorous and simple numerical laws which would bring order to the seeming chaos of the world of materials. These were the laws which provided the foundations for the great development of chemical science and technology in the nineteenth century, a development which in the magnitude of its accomplishment but also in the logical rigor of its adherence to canons of number and form, we regard as the classical period of chemistry.

It is possible now to look back and see that the classical chemistry of the later nineteenth century, the self-confident, productive science, positive in its assertions, successful in practice—the chemistry of the new industrial age, of metals and alloys, of ores and fuels, of dyestuffs, bleaches, synthetics, paints, of drugs, explosives, anesthetics, antiseptics, fertilizers—rested upon a base of only a few general principles. These laws of chemistry are not laws in the sense of the mathematically shaped laws of physics or physical chemistry—$PV = k$, for example. They are general assumptions at the base of the classical chemist's equations and experiments. They are comparable only to such laws of quantum mechanics as that which specifies that the principle quantum numbers shall be integers.

The first of these laws is the principle of Lavoisier, the law of conservation of matter (1790). The second is Richter's Rule: the elements combine to form compounds in fixed proportions—equivalent weights (1795). The third is the law of constant proportions: a chemical compound has a fixed composition by weight (1800). The fourth is the law of multiple proportions (1803): "If two or more compounds can be prepared out of the same elements, these compounds will be composed of simple multiples of the weights of the elements."

In his passion for universal measurement, Cavendish found in 1766 that different amounts of different bases were required to neutralize a fixed amount of a given acid. These different weights he called *equivalent.* Later reversing the process, he neutralized a fixed amount of potash with different amounts of sulphuric and nitric acids and then noted that the same ratio of weights of the two acids was required to complete the reaction with a fixed amount of marble. Tables of equivalent weights were drawn up by J. B. Richter (1762–1807) in the last decade of the eighteenth century. Richter, who had studied under Kant at Koenigsberg, was obsessed with the idea of the mathematization of nature. According to Kant, who was in this following the Pythagorean tradition, our concepts of space and time (and therefore of geometry and number) are a priori truths—intuitively known. We do not *learn* that one and one make two, that if equals be added to equals the results are equal. We know them in advance, intuitively, and therefore the mathematics that we deduce from these and the other simple axioms by the operations of logic—the Euclidean geometry, the algebraic theorems—are not simply free creations of the human mind or convenient codes which we have adopted through custom, but truth, absolute and immutable. Here in number and form, Kant argued, are the fixed unchanging absolutes that reason elsewhere denies. For Richter this philosophy led to the search for number and quantity in chemistry.

Richter used the term *stoichiometry* to define the science of measuring the elements—which he considered the central part of chemistry. "The elements," he wrote, "must have among themselves, a certain fixed proportion of mass." Or rephrasing his rule, it is possible to assign to every element a number such that the combining weights of the elements will be in the proportions of the numbers.

Richter's Rule reveals a series of mass-numbers (equivalent weights) through synthesis, the combining of chemicals. Proust's law of constant proportions reveals the same series of numbers in the analysis of compounds: "We must recognize that invisible hand which holds the balance for us in the formation of compounds." J. L. Proust (1754–1826), a French chemist brought to Madrid by Charles IV, based his law on his careful analyses of the several different oxides and sulfides that a single metal (copper was his principal example) may form. Using modern figures, the ratio of the weight of copper to oxygen in cuprous oxide (CuO) is nearly 4:1; in cuprite (Cu_2O) nearly 8:1; in covellite (CuS) the ratio of weights of copper to sulphur is 1.98:1; in chalcocite (Cu_2S), 3.96:1.

Berthollet attacked this concept, insisting that chemicals could react with each other in varying proportions. Copper heated in air would react with oxygen in a continuously increasing proportion up to a limit given by Proust's ratio. There were not two oxides of copper, Berthollet asserted, but an infinite series of oxides of copper, proved by the color changes of the metal with heating. Alloys like gold and silver, solutions, glasses were all capable of continually varying compositions between fixed limits.

Although Berthollet was mistaken in not distinguishing mixtures from compounds in the examples of the oxides and sulphides he chose, ordinary analyses continued to give widely disparate results, for reasons which were not to be understood until Mitscherlich's law of isomorphism was formulated in 1819. Berthollet's experience was strongly affected by the French Revolution of 1789, which was accompanied by a major breakdown in ordinary processes of manufacture and trade, while the armies of Europe moved in concert to attack the republic and restore the aristocrats. The problem of supplies for the new republican conscript armies, particularly saltpeter for gunpowder, was met by incorporating scientists into military and government service. There was nothing new about this; Charles II had made ample use of the talents of the Royal Society, and the Paris Academy since its founding had been at the service of the French monarchy. Lavoisier in 1775 had headed a Royal Commission for the improvement of the manufacture of saltpeter. Nevertheless, the scope of scientists' involvement with the revolutionary government was now greatly enlarged. The Commission of Weights and Measures was one outstanding example. From the first Berthollet was swept up in government service and devoted himself extensively to practical chemistry—the chemistry of processes on an industrial scale, which are by no means the same as they are in the laboratory. He worked on the processes of steel making; he introduced Scheele's chlorine as a bleach. With Macquer, he directed the French dye industry, and he studied the chemistry of soaps and oils.

Berthollet was one of the scientists and scholars taken by Napoleon on the Egyptian expedition. He concentrated upon the effects of the relative quantities of the reagents in a chemical reaction, and as a result was able to formulate the first of the mass action laws.

Berthollet's Pythagorean contemporaries read numbers into their experimentation even when numbers did not occur. "It appears . . . that oxygen joins to nitrous gas sometimes 1.7 to 1, and at other times 3.4 to 1," John Dalton (1766–1844) wrote in 1803 of a series of experiments which seem hardly likely to have yielded results so neat, but which were to have a profound influence on the evolution of science.

Dalton also came from that stock of English dissenters, hard-working weavers and tailors of the Midlands, which had produced such luminaries as Praisegod Barebones at the time of the Parliamentary Revolution, and which in the eighteenth century turned to learning and nature to find those orderly traces of God's handiwork denied them in the political sphere. Dalton's father was a poor weaver and the family belonged to the Society of Friends, in whose Quaker school at Eaglesfield Dalton studied and taught. He began with an interest in meteorology and a firm commitment to Newtonian mechanism which soon led him to results of the greatest import.

"Newton has demonstrated clearly, in the 23rd Prop. of Book 2 of the *Principia,* that an elastic fluid [gas] is constituted of small particles or atoms of matter, which repel each other by a force increasing in proportion as their distance diminishes," Dalton wrote, describing the origins of his thought. Following experiments in which he determined that the pressure of a fixed volume of gas was proportional to the temperature,* Dalton identified the repulsive force between atoms with the heat fluid which Guyton de Morveau had named caloric. He

* Previously observed by J.A.C. Charles (1746–1823), hence Charles' Law.

thought of a gas as consisting of hard, round, impenetrable spheres surrounded by atmospheres of caloric fluid. How would this fit the pressure of air—a mixture of at least three distinct gases—azote (nitrogen), by this time known to make up about four-fifths of the air by weight, oxygen, about one-fifth, and water vapor? Of the total pressure of the atmosphere, one-fifth is the partial pressure of the oxygen and four-fifths is due to the partial pressure of the nitrogen (neglecting the smaller contributions of other gases). Removing the oxygen by fixing it in a metallic calx or by some other means removes the contribution of the oxygen to the total pressure and reduces the pressure by one-fifth. The pressure exerted by a given quantity of a gas on the walls of its container is the same *regardless of the presence or absence of other gases.*

This observation of Dalton's, which is known as the law of partial pressures, appeared to contradict the Newtonian model of a gas. According to the Newtonian model, the force exerted by each particle of nitrogen depends upon the distances between the particles. The introduction of particles of oxygen would reduce the average distance between the particles of the gases and therefore increase the pressure exerted by the particles of the nitrogen in *addition* to the increase in pressure contributed by the oxygen. But the empirical law of partial pressures meant that the force exerted by a particle of nitrogen did not change with the addition or subtraction of oxygen. Dalton saved Newton's completely erroneous model by assuming that the gas particles exerted their pressures only upon atoms of the same gas and ignored atoms of other gases; in other words, the atoms of different gases were distinguishable, atoms of the same gas were alike. This difference, he came to believe, must be a difference of size.

> The different *sizes* of the particles of elastic fluids under like circumstances of temperature and pressure being once established, it became an object to determine the relative *sizes* and *weights,* together with the relative *number* of atoms in a given volume . . . Thus a train of investigation was laid for determining the *number* and *weight* of all chemical elementary principles which enter into any sort of combination one with another.

These ideas, including the concept of chemical combination between atoms, comprise Dalton's atomic theory. A compound gas such as water was made of binary particles consisting of one atom of hydrogen joined to one of oxygen. This led to the law of multiple proportions. Chemical combination occurs by the union of atoms in simple numerical ratios. Dalton represented the atoms by little round symbols and joined these in order to visualize compounds. The first compound of two elements, A and B, he assumed to be binary—AB—and therefore he represented the water molecule by two little circles, ⊙○, the first standing for the hydrogen atom. A second or third compound of two elements would be a ternary molecule, thus carbonic acid was represented by ○●○, a central carbon atom flanked by two oxygens, a fourth compound by a quaternary molecule, and so on. On Dalton's thirty-seventh birthday, in 1803, he entered in his notebook a table of atomic and molecular weights relative to hydrogen with a set of symbols for ten different gases.

Dalton's atomism is a model, a mechanical picture which explains beautifully the fundamental empirical laws of chemistry. Through it we understand the law of constant proportions of Proust. We understand the combining proportions of Richter's Rule—these are simply the relative molecular weights. Richter's tables of proportionate weights had been appended to Berthollet's chemistry textbooks and, according to Dalton's pupil W. C. Henry, were the source of Dalton's ideas on atomic weights. The law of multiple proportions is uniquely Dalton's, established by rather indefinite experiments on mixing nitric oxide with air over water, and by careful experiments in 1804 on marsh (methane) CH_4 and olefiant (ethylene) C_2H_4 gases; he found the weights of these to be, respectively, 6.3 and 5.3 times the weight of hydrogen: "If we reckon the carbon in each the same (4.3), then carburetted hydrogen [marsh] gas contains exactly twice as much hydrogen as olefiant gas does."

The published researches of Proust, Dalton realized, confirmed his law. The atomic hypothesis also explains Dalton's law of partial pressures, but not in the way that Dalton thought. He had confused Newton's assumption that the pres-

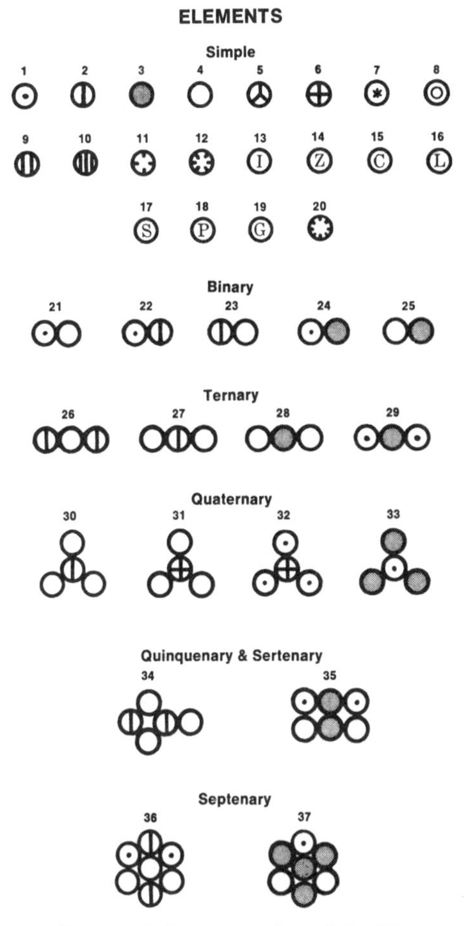

Figure 6. Dalton's Representation of the Elements (1–20) and Compound Molecules Which He Called "Atoms" (21–37)
The elements, or "ultimate particles," are arranged in the order of relative weights from hydrogen (1) through mercury (20), with a relative weight of 167. No. 4 is oxygen (relative weight 7). Binary "atom" 21 is "atom of water or steam."

sure of a gas depended upon a repulsive force exerted by the particles of the gas, with a proof that this was the true model of a gas. But Newton in the *Principia* was careful to distinguish between the induction of natural law and the mathematical deduction of the consequences of a principle. The *Principia* are the mathematical principles of natural philosophy. Books I and II contain only hypothetical laws: "*If* the density of a fluid which is made up of mutually repulsive particles [one of Boyle's two possible models for a gas] is proportional to the pressure, the forces between the particles are reciprocally proportional to the distance between their centers," Newton wrote, and proved by purely mathematical methods. "And vice versa, mutually repulsive particles, the forces between which are reciprocally proportional to the distance between their centers, will make up an elastic fluid the density of which is proportional to the pressure."

If we could make a synthetic fluid of peculiarly charged particles, repelling each other as Newton assumes, his theorem tells us it would obey Boyle's Law. "Whether elastic fluids do really consist of particles so repelling each other is a physical question," he added.

Daniel Bernouilli (1700–1782) had shown in 1738 that Boyle's alternative model of a gas made of colliding spheres would also account for the law. The gas particles would be of negligible size compared to the average distance between particles. No repulsive force attached to the particles; the gas exerted its pressure through the collisions of myriads of particles.

By 1808, J. L. Gay-Lussac (1778–1850) had found an important turn in the law of multiple proportions: the volumes of reacting gases were in simple multiple proportions. Cavendish had originally reported that hydrogen reacted with oxygen in the volume ratio of 2.02:1, and Lavoisier (1784) had reported with spurious precision that twelve volumes of oxygen reacted with 22.924345 volumes of hydrogen, a ratio of 1:1.9. Both experiments had started with two hydrogen to one oxygen. Even Dalton's first instance of multiple proportions had referred to the volumes of nitrogen oxides rather than their weights.

The law of integral volumes is a necessary conclusion from Boyle's Law. All gases obey Boyle's Law (approximately), and what is more remarkable, the constant of proportionality—the k in the equation $PV = k$—has the same value for all gases.

If Boyle's Law is coupled with the law of Charles and Dalton so that it reads PV = RT with P, V, and T representing pressure, volume, and temperature in appropriate units—then the law will hold *numerically* for the experiment performed with any of the ordinary gases, using the same value, R, for the constant of proportionality. Here is an instance of a number built into the frame of things as they are, a number which cannot be deduced from any a priori assumptions, a result not to be obtained from the introspection of a Descartes or the intuition of a Kant. The gas constant R, like the constant of gravitation, or the velocity of light, is apart from and outside of the human mind. It must remain forever a stumbling block for those philosophies of total idealism which deny objective existence to nature and separate the human mind from the natural world.

If we repeat Boyle's experiments for all the gases, always using the same pressures, volumes, and temperatures, there will emerge a set of numbers characterizing the proportionate weights of unit volumes of the separate gases—the equivalents of Cavendish. The gas laws are summarized in the form of the ideal equation, PV = nRT, with n standing for the number of equivalent weights or standard volumes under consideration.* The law of Gay-Lussac, which is a law of chemistry and yet contains an important physical implication, asserts that in any chemical reaction between gases the proportionate n's shall be simple integers. The nature which refused rationality to the Pythagoreans and, by confronting the Greeks with the irrational numbers, persuaded them to abandon arithmetic for geometry, here in Gay-Lussac's Law confirms again the integral character of matter. The simple integers appear in our science of matter, and the model for matter so characterized is atomistic.

We could dispense with models entirely, in the manner of Pierre Duhem and Ostwald, or in the positivism which led Lavoisier to relegate the question of atoms to metaphysics. We would have a science of pure number and the logical relation-

---

* The standard volume of a gas is the volume occupied by 1 gram molecular weight of the gas at 0° C. with a pressure of 1 atmosphere. It is 22.4 liters.

ships between numbers. Experiment would yield the numbers of nature and a set of fundamental relationships between these such as the ideal gas law above. Logic and mathematics would operate on these to produce the laws, the interrelationships governing all phenomena. We would attach no significance to the fact that again and again in the laboratory study of matter the numbers which appear are simple integers, 1, 2, 3, . . . n.

> So far as I know [says Sagredo in Galileo's *Dialogues on Two New Sciences*], no one has yet pointed out that the distances traversed, during equal intervals of time, by a body falling from rest, stand to one another in the same ratio as the odd numbers beginning with unity.

The Pythagorean implications of a material world based on integers were inescapable. In 1811 Amadeo Avogadro (1776–1856), professor at Turin, proposed that the simple integral ratios of the volumes of combining gases could depend only on the relative numbers of the combining molecules and the resulting compound molecules.

> There are also simple relationships between the volumes of gases and the number of the simple or compound molecules which form them. The first hypothesis which appears to apply and which also appears the only one admissible, is to suppose that the number of unit molecules in any gas whatsoever, is always the same for equal volumes, or is always proportional to the volume.

One volume of oxygen and two volumes of hydrogen yield two volumes of water vapor. We have Avogadro's number, N molecules of oxygen, reacting with 2 x N molecules of hydrogen to give 2 x N molecules of water. If we assume the molecules of water to be ternary—that is, to contain two molecules of hydrogen and one of oxygen—we should have N triple molecules. Since one volume plus two volumes yields not one but two volumes, not N but 2N triple molecules, the model should be modified. Avogadro proposed that the molecules of oxygen and hydrogen were themselves binary, so that the single volume of oxygen consisted of N molecules or 2N atoms; the two volumes of hydrogen contained 2N molecules or 4N

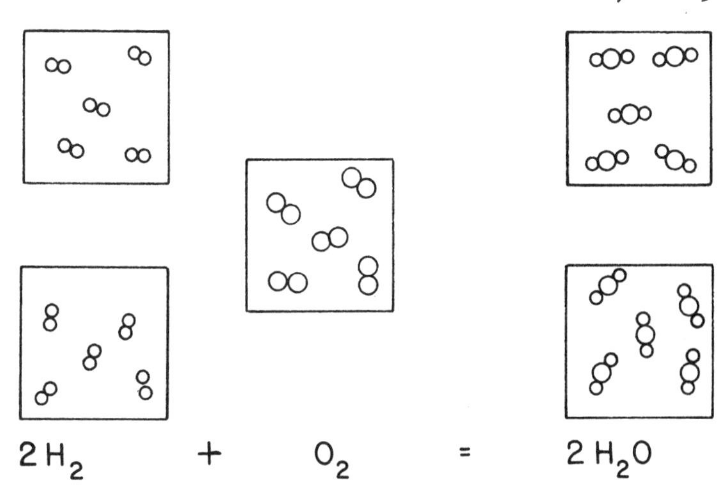

Figure 7. Avogadro's Number
Under the same conditions of pressure and temperature, equal volumes of any gas whatsoever contain equal numbers of molecules.

atoms; and the two volumes of water vapor contained 2 x N ternary molecules or 6N atoms.

Dalton, wedded to the caloric concept, wished to account for his law of partial pressures by different sizes for his atoms to keep them at different distances. The sizes were proportional to the differential attraction for caloric of the atoms. But in this case, Avogadro argued, the volumes of equivalent weights would be different. Only if the molecules are far enough from each other—"with invariable distance between molecules"—can Gay-Lussac's simple relationships between volumes be understood.

Avogadro's hypothesis was not accepted by either Dalton or Gay-Lussac. In fact it would have been little more than a frill tacked on to Daltonian atomism, except for its immediate usefulness in correcting atomic and molecular weights. If the number of molecules is the same for equal volumes of different gases, the molecular weights (Avogadro used "mass proportions") would be simply the proportionate densities or weights of equal volumes. From the densities Avogadro obtained 15.074

for the weight of oxygen relative to hydrogen; 13.238 for nitrogen (now 16 and 14 respectively). Dalton originally gave 5.5 and 4.2; with the improved figures available to Avogadro he would have used exactly half of Avogadro's weights, since he gave the composition of water as HO and of ammonia as NH. But, Avogadro wrote, ". . . as one knows that the ratio of the volumes of hydrogen to oxygen in the formation of water is that of 2 to 1, it follows that water results from the union of each molecule of oxygen with two molecules of hydrogen."

Similarly, ammonia must be NH_3 and the three oxides of nitrogen N_2O, NO, and NO_2. To explain the fact that two volumes of water are obtained from the union of two hydrogen with one oxygen, Avogadro proposed that the molecules of elementary oxygen and hydrogen (and nitrogen) must be binary; thus (in modern notation):

$$2H_2 + O_2 = 2H_2O$$

Using his concept as a powerful tool for the determination of molecular weights, Avogadro in his earliest paper established weights for carbon from carbon dioxide and began the process of unraveling the molecular formulae of metallic oxides. Discrepancies in published weights of sulphuric acid led him to suggest correctly that hydrogen entered into the acid molecule.

Although this idea was taken up in a distorted and erroneous form by Ampère, Avogadro could point out in 1814 that neither Gay-Lussac, Davy, nor Berzelius* had employed it. Davy used Avogadro's evaluation of proportionate weights but expressed them as the least proportion in which bodies enter into combination, ". . . inasmuch as he has not made use of my hypothesis on the constitution of gas . . . ," Avogadro wrote.

Berzelius worked directly from Gay-Lussac's law to obtain similar results but without Avogadro's reasoning; he did not use the concept of binary molecule for elementary gases. Nevertheless, in 1814, 1818, and 1826 Berzelius published a series of increasingly accurate tables of atomic and molecular weights relative to the weight of oxygen taken as 100, and to

* Jöns Jakob Berzelius (1779–1848) was educated as a physician at Uppsala and became the foremost chemist of the age.

hydrogen taken as 1. Except for silica, potash, soda, and silver oxide, his formulae in 1826 were correct and his weights nearly those of modern measurements.

In 1819 the French chemist Pierre Dulong (1785–1838) and the physicist Alexis Petit, of the École Polytechnique, had determined a most remarkable law—that the heat capacities of the elementary solids were all the same when referred to equivalent weights (six calories per gram molecular weight per degree Centigrade). Berzelius used the law to correct his formulae for such compounds as As_2O_3, As_2O_5, Hg_2O, CuO, which he had originally written as AsO_3, AsO_5, HgO, CuO_2. His early values for the atomic weights of these elements were just twice the values which would bring agreement with the law of Dulong and Petit, and he therefore doubled the proportion of metal in the oxides and halved the atomic weights. In the cases of soda, potash, and silver he had assigned formulae of RO_2 in 1814 and 1818. His method yielded formulae RO in 1826, with heat capacities for the metals R still double those of the ordinary metals. The formulae should be R_2O.

The law of Dulong and Petit was purely empirical—there was no explanation, no hypothetical model to account for it until the kinetic-molecular theory appeared half a century later. There were, besides, in ordinary experiments, departures from the law of up to 20 percent. Chemists preferred to think that the disagreement between the law and the formulae for soda, potash, and similar strong bases showed the weakness of the law rather than the error in the formulae. By 1821 Avogadro's hypothesis was widely enough adopted for him to note that it was in general use with no credit to its author. The French chemist Dumas was completely converted until he measured the vapor densities of sulphur, phosphorus, arsenic, and mercury. The molecules are S_6, P_4, As_4, and Hg, and believing that such extreme values must be impossible, Dumas abandoned the Avogadro hypothesis. As late as 1866 Josef Loschmidt (1821–1895), who has been credited with the first determination of the absolute value of the Avogadro number, wrote after his calculations, "Such a conclusion . . . that 1000 H_2 contained in hydrogen gas should also exactly [correspond to] 1000 valerian amyl acid or chlorbenzoyl molecules contained in the appropriate gas . . . goes very strongly against all probability."

It was the nascent positivism of the age which prevented chemists from seizing to the full the brilliant hypotheses with which they were presented. In full reaction against the Pythagoreanism which had made the seventeenth century so successful, the early-nineteenth-century chemists abandoned atomic weights for the security of equivalent weights. "The first duty of a scientist is to be precise and accurate" is the shibboleth of small minds unwilling to desert the safety of the balance and the test tube for the stormy adventures of the mind. It is to confuse the ethical obligation of honesty in experimentation with abdication of the creative mind. Lavoisier, the apostle of this protopositivism, would have abandoned his theory of oxidation had he not placed more confidence in the assumption of the conservation of matter than in the experiments in which he did not even weigh the reactants. Lavoisier calculated the volume of hydrogen reacting with oxygen to six decimal places for an experiment which was in error by nearly 10 percent.

It is difficult to avoid characterizing the failure to adopt Avogadro's hypothesis by the triumphant chemistry of the early nineteenth century as a failure of the imagination. "Theoretical physics has become a perpetual challenge to the integrity of the accurate mind," Pierre Duhem wrote in 1905:

> For the devil has not only touched the texts and courses intended for future engineers. It has penetrated everywhere, propagated by the hatreds and prejudices of the multitude of people who confuse science with industry, who, seeing the dusty, smoky, and smelly automobile, regard it as the triumphal chariot of the human spirit. . . . We no longer think of putting into young minds ideas and principles, but substitute numbers and facts.

By 1860 the situation had become critical. "It was apparently considered a sign of independence of thought for every chemist to have his own set of formulae," Partington asserts, adding that Kekulé gave nineteen different formulae for acetic acid in 1861. Berzelius had introduced a system of chemical notation based on the first letters of the Latin names of the elements following Thomas Thomson's suggestions of 1802. These had been extensively modified by Leopold Gmelin (1788–1853), whose tables of equivalent weights had largely supplanted the atomic weights of Berzelius. Gerhardt in 1843 had judiciously corrected Berzelius' weights for such strong bases as soda and potash by halving them, but in the process halved *all* Berzelius' atomic weights for metals. Kekulé and Wurtz used Berzelius' symbols but Gerhardt's weights. There appeared to be two chemistries, organic and inorganic, with atomic weights which changed according to which chemistry was used.

An international congress at Karlsruhe in 1860 failed to bring about any agreement, but it was at this congress that Stanislao Cannizzaro (1826–1910) showed that the clue of agreement was in Avogadro's neglected hypothesis. Superficially, this agreement was between the results of organic and inorganic chemistry. Avogadro's hypothesis was a model of the fine structure of matter. Beneath the level of direct observation, smaller than the wave length of the light which we must use for microscopic observation, smaller than the least particles which can be distinguished through the electron microscope, lies the atom. There is not one ultimate atom, but according to Dalton in 1803, there are as many kinds of atoms as there are elements.

The basic concepts of classical chemistry were all established between the years 1790 (Richter) and 1811 (Avogadro): The particles of a gas (molecules) are made of one or more atoms. All the particles of the same gas are alike. Under the same physical conditions (temperature, pressure, etc.) the same volume of any gas contains the same number of molecules. Just as Dalton's hypothesis was the single idea connecting the two empirical laws, Richter's Rule and the law of multiple proportions, Avogadro's hypothesis connected these with Gay-Lussac's law of integral volumes. Dalton found his law of multiple proportions after the hypothesis of atomism; Avogadro, his hypothesis after the law.

It is possible to have the empirical laws without the hypothesis, to reconcile the differing atomic weights and formulae of organic and inorganic chemistry without a word of atoms or particles, or without a commitment to a particular model of the fine structure of matter. One can say of the Avogadro number that it too is a universal constant, a pure number, devoid of any physical significance. It is a constant (N) of the ideal gas law: $PV = nkNT$. Or, assuming that all the equations of organic and inorganic chemistry are written out, it will be possible to assign to each element a unique weight number a (the atomic weight), which multiplied by a constant N (the Avogadro number) gives unit combining weight for the element.

The definitions of each of our numbers may in fact be written without any resort to the hypothesis, without any mention of atom or molecule or the fine structure of matter, but purely in mathematical terms—constants, integers, numbers, and geometric forms built into the frame of the universe. But how difficult a task it is to separate the number from the entanglement of the hypothesis; how awkward the expression above. With hindsight, having the number and making free use of it to determine the wave lengths of X-rays, the separation between atoms in a crystal, and the charge on the electron, we can as an exercise divorce our constant from any hypothetical meaning. But how simply and clearly it emerges from Avogadro's statement. It is the number of molecules in unit volume of a gas under standard conditions. To find the number we resort to mathematical analysis of the random (Brownian) motion of particles in a fluid. How difficult if not impossible it would be to devise such an experiment, such a method, without the guiding thread of the kinetic-molecular hypothesis.

> One single number has more real and permanent value than an expensive library of hypotheses; the attempt to penetrate by hypotheses to the inner recesses of the world order is of a piece with the efforts of the alchemists.*

Yet it was not by such reasoning that the Avogadro number was conceived or found, or the muddle of chemical weights and formulae resolved. There can be such a thing as oversophistication, and Avogadro's straightforward reasoning was in the end more productive than refinements of analysis. His-

* J. R. Mayer, quoted by Mason, p. 499.

torically it was, after all, by the piling of hypothesis upon hypothesis that first Dalton and then Avogadro established the atomic-molecular hypothesis—a structure so useful, so fertile, withal, so beautiful, that to question the legitimacy of its origins seems sheer perversity.

8

The Development of Organic Chemistry

WHY NATURE CANNOT GIVE THE POWER OF MOVEMENT TO ANIMALS WITHOUT MECHANICAL INSTRUMENTS, AS IS SHOWN BY ME IN THIS BOOK ON THE WORKS OF MOVEMENT WHICH NATURE HAS CREATED IN THE ANIMALS. . . . AND WE SHALL BEGIN BY STATING THAT EVERY LOCAL INSENSIBLE MOVEMENT IS PRODUCED BY A SENSIBLE MOVER, JUST AS IN A CLOCK THE COUNTERPOISE IS RAISED UP BY MAN WHO IS ITS MOVER.

—LEONARDO

In 1832 Wöhler and Liebig wrote of the "dark forest of organic chemistry," a trackless waste of amorphous complexity. If much of the mineral kingdom fell neatly into the simple numerical pigeonholes of mechanistic law, the material stuff of the living world seemed insuperably complex. The essential characteristic of life, continual change, appeared to preclude any fitting to mechanism or classical laws. Nevertheless, the middle of the nineteenth century saw this stubborn thicket reduced to a beautifully ordered garden. The study of living matter proved in no way less complex than had been feared, but far from being formless, these materials of life turned out to be patterned in elaborate and intricate ways, yielding, however, perfectly simple additions to complete the classical laws. What had appeared to be amorphous complexity became richness of design. The simple geometry of the crystal of the inorganic chemists became the simple geometry of the carbon atom of the organic chemists. This ultimate resolution could not have been predicted by Wöhler and Liebig in 1832, but their confidence in a solution, and their trailblazing through the forest, made them in fact the patron saints of mechanism in the nineteenth-century development of organic chemistry.

The doctrine of vitalism, the philosophic counterpoint to mechanism, was to reach its climax in the romantic movement of the same century, especially in Germany, as it was exemplified by Goethe. The idea that the human mind as an expression of the spirit of life could never be encompassed in an equation or a test tube—an earlier version of our resistance to being reduced to numbers on a punch card—was beneath the vitalist refusal to extend the new science of the eighteenth century to the animal and vegetable kingdoms. The parochialism which a century before had seen man at the center of the universe, the pivot about which all things revolved and the purpose for which the universe had been created, took refuge now in vitalism. There was a fundamental difference between the organic and the inorganic kingdoms. There was a vital spirit which animated and made for procreation and growth. Incinerated, the stuff of living things reduced to carbon, nitrogen, ash, phosphorus, and water; to the familiar elements of the inorganic kingdom. But the vitalists held that no inorganic process, no laboratory process could ever produce an organic compound.

These ideas provided the resistance against which the mechanistic chemists of the nineteenth century sharpened their minds and developed the "dualist" and "unitary" concepts, as well as theories of radicals, of substitution, and of types. They also provided the philosophical background out of which grew the theory of evolution (1859), historical and therefore antimechanical, as well as the germ theories of fermentation and of disease. On the one hand a geometric and therefore mechanistic impulse inspired the researches which put structure and pattern into organic chemistry and gave it dimensions, creating a science of chemistry in three-dimensional space—stereochemistry. Yet on the other, the same research led Pasteur to a brilliant and fruitful career which was to make a modified vitalism again respectable.

In its earlier form, as expressed by Paracelsus in the sixteenth century, vitalism was the ancient Greek concept of the world as organism, a concept first seriously challenged by Cartesian dualism. Even Leonardo saw no contradiction in his program—"We shall describe this mechanical structure of man . . ."—and the organic view of the world:

> The water which rises in the mountains is the blood which keeps the mountain in life. If one of its veins be open either internally or at the side, nature, which assists its organisms, abounding in increased desire to overcome the scarcity of moisture thus poured out is prodigal there in diligent aid, as also happens with the place at which a man has received a blow.

For Paracelsus growth was the fundamental natural process; all nature was alive. The fundamental matter-stuff of Creation nourished eternal seeds which waxed and waned, growing, dying, and beginning again, guided by internal spirits or vital forces. Everything, a rock, a river, a tree, a mountain, had its guiding power—its *archeus,* Paracelsus called it. It was a formative force or virtue which Kepler in the seventeenth century called "plastic spirit" and which he asserted was learned in the whole of geometry and was responsible for the growth of crystals in the rocks. In the twentieth-century mines of highland Peru the Indian miners always contrive to leave some ore behind in a worked-out stope as seed, to grow into ores again for future miners to harvest. Accustomed to seeing stalactite-like precipitates from the mine waters, they see no contradiction in extending the conservation practices of the good hunter or fisherman to the interior of the mountains. They ignore the Cartesian distinction between the mineral and organic kingdoms.

The iatrochemistry of Paracelsus, elaborated by van Helmont and by the visionary Jakob Boehme (1575–1624), inspired Gottfried Wilhelm Leibniz (1646–1716) to a doctrine of substance which influenced the German nature-philosophy movement and offered, until modern times, the only serious challenge to mechanism. In the Leibniz conception, the *archei* of van Helmont became units of living force, *monads,* ". . . the very atoms of nature—in a word, the elements of things."

The monads were point centers of living force, differing from the atoms of the mechanists, which were inert and material. By their aggregation and motions they accounted for the properties which mechanism would have placed in matter. Having no parts, extension, or figure, being continuous spiritual entities acting through a kind of inertia or self-power, completely gradational like the matter of the Cartesians, Leibniz' monads were free of the gross logical difficulties of early atomism.

While eighteenth-century England and France refined the precise and elegant Baroque materialism which saw the universe as an ornamental clockworks with God as clockmaker, in Germany the more romantic archei and monads dominated thought. The view that the mechanical laws of chemistry could never be extended to living things persisted long after the rise of organic chemistry. Among the chemists themselves there was a widespread belief that while analysis was possible, the synthesis of organic compounds could be accomplished only by the vital force. The biologist George Wald has pointed out that it is hardly reasonable to allow a plant or a worm to synthesize a fiber, but deny the same power to a man.

In his treatise of 1789 Lavoisier, the Compleat Mechanist, foresaw no division between inorganic and organic chemistry. His ignorance of the complexities of organic chemistry protected him from pessimism as he wrote that:

> . . . most of the vegetable acids . . . have radicals composed of hydrogen and charcoal, combined in such a way as to form single bases, and that these acids only differ from each other by the proportions in which these substances enter into the composition . . . the radicals from the animal kingdom and even some of those from vegetables, are of a more compound nature, and, besides hydrogen and charcoal . . . they often contain azote [nitrogen] and sometimes phosphorous . . .

His concept of the organic radical was adopted by Berzelius, who was then still dividing chemistry into scarcely related animal, vegetable, and mineral categories. Berzelius undertook a series of analyses of sugar, starch, and organic acids (citric and tartaric) which established that Richter's rule of equivalents and the law of multiple proportions extended to the organic kingdoms. He combined these discoveries with the ideas of Lavoisier into his electrochemical or "dualist" theory of the formation of compounds. Following Lavoisier, Berzelius thought acids to be dual combinations of radicals with oxygen. Bases, he showed, were dual combinations of metals with oxygen, and salts were combinations of acids with bases. Electrolysis revealed a polarity in decomposing compounds which he believed accounted for the dualism. Ultimately, Berzelius contended, all compounds would be shown to be a union of two oppositely charged integral units—radicals which, like the elements of eighteenth-century chemistry, retained their identity in chemical reactions. They could enter into combination and be regained again from combination in their original form.

But the new electrochemistry did not seem particularly relevant to organic chemistry, and it was Berzelius himself, as late as the 1831 edition of his *Treatise of Chemistry,* who expressed the widely held belief in the requirement of a vital force for the preparation of organic compounds, a belief hardly disturbed by Wöhler's demonstration in 1828 of a simple synthesis of urea. As Wöhler wrote in 1828,

> . . . I obtained the unexpected result that by the combination of cyanic acid with ammonia, urea is produced: a fact all the more remarkable in that it offers an example of the artificial formation of an organic material, and even of animal nature, by means of inorganic principles, . . .

Friedrich Wöhler (1800–1882) studied in his youth with Gmelin at Heidelberg and Berzelius at Stockholm. At the age of twenty-three he analyzed cyanic acid but in the following year the twenty-one-year-old Justus von Liebig (1803–1873) obtained an identical analysis for fulminic acid, a compound distinctly different in its properties. When he was convinced that both analyses were accurate, Berzelius proposed in 1827 as a general phenomenon that ". . . the simple atoms of which substances are composed may be united with each other in different ways." To this phenomenon of distinct chemical species with the same composition he gave the term *isomerism.** Led by this to the examination of other isomers, Wöhler heated ammonium cyanate NH_3CNOH and obtained the isomer urea $(NH_2)_2CO$. The cyanate could be prepared from purely inorganic materials by reacting ammonia and carbon (graphite) and alkali (Scheele, 1782). In this way urea, which for centuries was obtained by alchemists from urine, was duplicated by synthesis. Synthetic chemistry was extended to the organic kingdom.

The vitalist position was stated as recently as 1947 by Lecomte du Noüy, who wrote that if the ingredients of a protein molecule were shaken at the rate of five hundred trillion times a second in a volume equal to the whole earth, it would take 10^{243} billion years to form a single protein molecule. Drowned in the flood tide of an organic chemistry that had synthesized drugs, fibers, sugars, fats, plastics, fuels, in limitless variety, the neovitalists now maintained that the most complex of all the organic compounds, the proteins—without which even the simplest of life forms is inconceivable—could not be produced by inorganic process. But even at the time du Noüy was writing, as the latest spokesman of the *ancien régime,*† scientists such as Oparin, Shapley, and Urey were trying to infer the physico-chemical conditions under which life had originated.

Wöhler had obtained organic matter from a meteorite in 1857, and Mendeléev accounted for the synthesis of hydrocarbons by the action of water on metallic carbides, which are commonly found in meteorites, postulated to make up much of the earth. Working in this tradition, H. C. Urey (1893–) transferred his attentions after the Second World War to the field of geochemistry. Urey speculated on the hydrogen

* *Allotropy* is usually taken to mean isomerism in the case of elements, and crystallographers use the term *polymorphism* to describe separate mineral species with the same formula. Berzelius referred to the existence of ethylene C_2H_4 and butylene C_4H_8, which he knew only as gases having the same composition but densities in the ratio of 1:2, as an example of *polymerism.*

† Something called ". . . the behavioristic, materialistic interpretation of the world, given 'scientistic' sustenance by the works of Darwin, Marx, Freud, Pavlov and others," was deplored in the *New York Times Magazine,* Aug. 16, 1963.

of interstellar gas and the methane and ammonia in the atmospheres of extraterrestrial planets. S. L. Miller, a student of Urey's at Chicago, who may have been influenced by the successful photosyntheses of formaldehyde and acetic acids by Melvin Calvin in 1951, passed a spark discharge through water vapor, hydrogen, ammonia, and methane and obtained a mixture of organic compounds including the amino acids, glycine, and alanine, subassemblies of the protein molecule.

Du Noüy's pessimistic statistics were mathematically correct but metaphysically inappropriate. He wrote of inorganic or mechanical processes but calculated something he called "chance."

Let molecules of hydrogen encounter molecules of oxygen and it is not by chance that water will form. In fact random processes will assure the overcoming of any initial hesitation of the elements. It is not by chance that the molecules of water are aggregated in lacy flakes of hexagonal symmetry; it is because this arrangement has a lower energy than other arrangements under the conditions of formation. In this arrangement the system of molecules occupies a sink in a free energy surface, and random processes serve only to assure that the molecules at some time find themselves on the slope leading down to the preferred arrangement. The energies that are associated with different patterns of arrangement of molecules are different. A change in position or arrangement of any part produces a change in energy. Since the configurations are slightly changed by random processes, such as vibrations due to heat or shock, those with lower energies are favored at the expense of those with higher. In a large and irregular valley of energy there are low spots in which methane, ammonia, water vapor, and hydrogen accumulate. There are other basins with high rims, like the craters of volcanoes, for formaldehyde and other organic compounds. Thermions of heat or photons of ultraviolet light collide with the molecules and knock them upward out of their energy basins only to have them fall back again on the slopes and roll back in. Some of them go over the rims of high energy around the sinks for amino acids, and if these sinks are deeper than the basins for the constituent

Figure 8. Dalton's Arrangement for the Atoms in a Snowflake, 1808

molecules, then the synthesis proceeds. Nor is it the mind of greater piety which requires the Deity to fashion together in a particular and arbitrary pattern the first molecules of protein, rather than to devise a universe from the beginning with such fundamental percipient monads that from them all else flows naturally.

But in jumping to the protein syntheses of the mid-twentieth century we are bypassing more than a hundred years of basic developments. The enormous complexity of organic materials meant for the chemists of the early nineteenth century that their chemistry could be understood only in terms of synthesizing concepts such as that of the radical in dualist compounds, as proposed by Lavoisier and Berzelius.

"... That which they have called *element,* or undecomposable substance, has been considered as such only with regard to the state of acquired experience. . . ." Dumas and Liebig wrote in 1837. "Actually to produce with three or four elements, combinations as varied as and perhaps more varied than those which form the mineral kingdom, nature has taken a course as simple as it was unexpected; for with the elements she has made compounds which manifest all the properties of elementary substances themselves. . . ."

The first of these radicals was cyanogen, CN, made by Gay-Lussac in 1815. There is

> . . . a very great analogy between prussic acid and muriatic and hydriodic acids [HCN, HCl, HI]. Like them it contains half its volume of hydrogen; and, like them it contains a radical which combines with the potassium, and forms a compound [KCN] quite analogous to the chloride and iodide of potassium. The only difference is that this radical is compound, while those of the chloride and iodide are simple.

It was Gay-Lussac, using his law of simple combining volumes, who showed that ethylene (C_2H_4) behaved like an autonomous unit or *radical,* combining with one part water to form alcohol, or two parts ethylene would combine with one part water to form ether:

alcohol		water		ethylene
C_2H_6O	=	H_2O	+	C_2H_4

ether		water		ethylene		ethylene
$C_4H_{10}O$	=	H_2O	+	C_2H_4	+	C_2H_4

Dumas and Boullay (1828) compared the behavior of ethylene (which was still known as olefiant gas) to ammonia (NH_3):

> In hydrochloric and hydroiodic ethers, one volume of gaseous acid is saturated by one volume of olefiant gas, just as in the neutral chlorides and iodides of ammonia, the acid and base being combined volume for volume. . . .
>
> One atom [sic] of nitric, acetic, benzoic, or oxalic acids, saturates four volumes of ammonia: now in the esters formed by these acids, one atom of each of them also saturates exactly four volumes of olefiant gas . . .

The esters are the aromatic compounds which are the scent and flavors of flowers and fruit, earlier prepared by Scheele (1782) by the reaction of the acids with alcohol. Berzelius saw in this work confirmation of his dualist views. The olefiant gas of the alcohol behaved as a unit, a radical, entering into combination with other units of opposite charge. He called the olefiant radical *aetherin* in a letter appended to a critical paper by Wöhler and Liebig (1832).

> When in the dark province of organic nature, we succeed in finding a light point, appearing to be one of these inlets whereby we may attain to the examination and investigation of this province, then we have reason to congratulate ourselves . . .

The point of light which Wöhler and Liebig found was a radical in the oil of bitter almonds to which they gave the name *benzöyl* and assigned the complex composition $C_{14}H_{10}O_2$. They showed that the oil of bitter almonds was a compound of the radical with hydrogen. It could be converted into a chlorbenzoyl by reaction with chlorine gas replacing the hydrogen. Benzoic acid was a compound of the radical with oxygen, which, like the hydrogen of the oil, could be "replaced by chlorine, bromine, iodine, sulphur or cyanogen, and the bodies proceeding thence, comparable with the corresponding compounds of phosphorus. . . ."

"We may well view [these results] as the dawning of a new day in vegetable chemistry," Berzelius wrote enthusiastically. He wrote the new findings with the symbol Bz for the benzoyl radical:

$\dot{\mathrm{B}}\mathrm{z}$ = Benzöylic acid [the dot standing for oxygen]
BzH = Bitter almond oil
BzCl = Chlorbenzöyl
BzS = Sulphuret of benzöyl
Setting Amid = NH_2
Bz + NH_2 = Benzamid
$\ddot{\mathrm{C}}$ + NH_2 = Oxamid
K + NH_2 = Potassiumamid

Berzelius set etherin (ethylene, to which he gave the formula C_4H_8) equal to E, and wrote formulae as dual compounds with the etherin radical:

E + ~~HCl~~ = Muriatic ether [hydrochloric ether; our ethyl chloride]
E + $\dot{\mathrm{B}}\mathrm{z}$~~$\dot{\mathrm{H}}$~~ = Benzöylic ether

The bar is Berzelius' symbol for two atoms—~~$\dot{\mathrm{H}}$~~ for our H_2O.

The fragrant muriatic ether had been prepared by Thénard (Louis Jacques, 1777–1857) in 1807 by the action of HCl on alcohol. Liebig rewrote the muriatic ether formula as the union of an alcohol radical *ethyl* C_4H_{10} with Cl; i.e., $C_4H_{10} \cdot Cl_2$ ethyl chloride, instead of the etherin radical C_4H_8 with H_2Cl_2 (Liebig's notation). Ether (C_4H_{10}) O was represented as ethyl oxide, EO; and alcohol as ether plus water, EO + H_2O. Liebig's arrangements do not fit Gay-Lussac's law of combining volumes, but they set the example which led to the identification of another radical, methyl (CH_3), represented by M, and another alcohol, methyl alcohol, written as methyl oxide plus water, MO + H_2O, or CH_4O, by Dumas and Peligot. In modern notation:

ethyl alcohol is $C_2H_5 \cdot OH$
methyl alcohol is $CH_3 \cdot OH$
glycerine is $C_3H_5 \cdot (OH)_3$; etc.

The reaction of each of the alcohols with acids gives rise to a series of esters.

In this way, by the fourth decade of the nineteenth century, chemists were ready to accept Dumas' and Liebig's definition of organic chemistry as a chemistry in which identifiable groups of elements in fixed proportions, the radicals, entered into compounds with other radicals and elements. In atomistic terms, groups of atoms retaining their identity and proportions through a series of compounds would form a recognizable class of particle intermediate between the atom and the molecule. This intermediate chemical building block was the radical. Organic radicals were compound, while inorganic radicals were simple. The discovery of the noxious *cacodyl* radical, $C_4H_{12}As_2$, by Robert Bunsen (1811–1899) served to confirm the basic concept even as new discoveries began to weaken Berzelius' simple dualism. The most important of these developments was the idea of substitution.

August Laurent (1808–1853) proposed to account for the transformations between organic compounds by concentrating on the replacements or substitutions of individual atoms of a fundamental structural unit, the nucleus. "Let us recollect that chlorine, iodine, bromine, fluorine, the nitric residue X, and the nitrous residue Y, may be substituted for hydrogen, and to a certain extent, fulfill its functions," he wrote in 1854.

The concept of substitution did not contradict the results of the radical-dualist theory so much as it changed the way in which chemists looked at these results and grouped them. "To arrange organic compounds in series, that is, to determine the laws according to which the properties in a given type are modified by substitution of an element or group of elements for other elements, this is the constant purpose of the chemist philosopher," wrote Charles Gerhardt (1816–1856). Not only did this *unitary* theory propose that the strongly electropositive hydrogen could be replaced by the strongly electronegative chlorine ion (to the dismay of Berzelius), but it also meant that radicals need not retain their identity in chemical reaction but could be altered by substitution of parts. Berzelius clung to his dualist concept and spent his last years trying to rewrite the formula of the substitution compounds in such a way as to identify and isolate opposite parts of invariable radicals.

After Berzelius and Gay-Lussac, Jean Baptiste André Dumas (1800–1884) was the principal experimental and theoretical supporter of the dualist-radical theory. But in 1834 he accounted for the action of chlorine on alcohol by proposing as a general law of nature that "chlorine possesses the singular power of removing the hydrogen of certain bodies, replacing it atom for atom." He called this the law of *metalepsy* and generalized it to include bromine, iodine, and "a half atom of oxygen."

Laurent stressed the persistence of the chemical properties in the compound after the substitution of chlorine for hydrogen; Dumas called these chemically related groups persistent *types*. In a parody, Wöhler, signing himself S.C.H. Windler (swindler), wrote to Liebig's journal that he had replaced all the atoms of manganous acetate with chlorine, winding up with $Cl_2Cl_2Cl_8Cl_6Cl_6$—an acetate making bleached fabrics consisting entirely of spun chlorine, which he said were widely sold in London.

Laurent's life was short and embittered by the scorn of his colleagues, even including Dumas, whose student he had been. But in seeking to account for the persistence of his substitution compounds, Laurent, who has been characterized as the most brilliant chemist of his day, proposed that the chlorine or hydrogen was linked to a stable grouping which was a radical or nucleus in the shape of a prism. In this way Laurent set in motion the train of ideas which was to reintroduce geometry into chemistry—another of the Platonic concepts too hastily discarded with the migration from alchemy.

Dumas' and Laurent's concepts of types followed naturally from the idea of substitution. In the middle of the nineteenth century it led directly to the organization of organic compounds into clearly recognizable categories—the types. From these types came, first, formulae which would express composition in terms of groupings of elements and radicals, and then the idea of a true structural chemistry, a chemistry in space. The recognition that key elements within the types always played the same role (the fourfold nature of carbon is the best example) led to new laws of classical chemistry. The law of valence opened the door to the concept of the chemical bond or the forces between atoms, a subject destined to become almost an independent branch of science in the twentieth century.

The concept of types was almost completely developed in the

short span of the years 1849 to 1857 by a handful of chemists in England, France, and Germany.

C. A. Wurtz (1817–1884) wrote in 1849 that ethyl and methyl amine might be viewed "as ammonia in which one equivalent of hydrogen is replaced by methylium C^2H^3 or ethylium, C^4H^5." A. W. von Hofmann (1818–1892) generalized this to show that all three of the hydrogens of ammonia could be replaced by organic radicals: ammonia NH_3, ethyl amine $NH_2 \cdot E$, diethyl amine $NH \cdot 2E$, triethyl amine $N \cdot 3E$, ethyl aniline $NH \cdot E\ An$ ($An = C_6H_5$), and so on. It was a new way of ordering the complexity of organic nature which immediately opened avenues of research. At the same time A. W. Williamson (1824–1904) elaborated with great success Laurent's idea of a "water type," expressing the formulae of alcohol, ether, and a host of related compounds as fundamentally that of water $\begin{matrix}H\\H\end{matrix}O$. "Alcohol is therefore water in which half the hydrogen is replaced by carburetted hydrogen, and aether is water in which both atoms of hydrogen are replaced by carburetted hydrogen: Thus,

$$\begin{matrix} H & & C^2H^5 & & C^2H^5 & \\ & O, & & O, & & O\ldots\text{''} \\ H & & H & & C^2H^5 & \end{matrix}$$

Williamson wrote acetic acid as $\left[\begin{matrix}C_2H_3O & \\ & O \\ H & \end{matrix}\right]$ and nitric acid as $\left[\begin{matrix}(NO_2) & \\ & O \\ H & \end{matrix}\right]$. He even undertook to rewrite Wurtz' amines as water types.

This kind of classification, with its almost endless possibilities of substitution and replacement, offered a powerful tool for the prediction and discovery of new compounds whose properties could be approximated in advance from the properties of related compounds in the same type. The possibility of a kind of engineering chemistry was uncovered, and those with foresight might have glimpsed a future world of synthetic materials, drugs, fuels, dyes, a world that looks back to the first syntheses of urea and acetic acid in much the same way as it looks to the first attempts by Peking man to break a stone into the desired shape instead of merely hunting for one which by chance would fit.

In 1853 Gerhardt proposed that four inorganic types—H_2O, HCl, NH_3, and H_2—among them would yield all organic compounds. Kekulé in 1857 proposed a marsh gas type CH_4, in which the knowledgeable student of chemistry will see a recognition of the principle of *valence,* the true basis of organic chemistry, the quadrivalent, tetrahedrally shaped atom of carbon.

The law of multiple proportions and the law of integral combining volumes were ready for a significant modification. An enormous number of chemical compounds and substitutions were known, and the chemists had a ready familiarity with the possibilities of substitution which led to the idea of valence.

Not only was there a weight number, proportional to hydrogen the *equivalent,* associated with every element, but there was a proportion number or number of equivalents. The elements form compounds with a certain number of equivalents of other elements. Edward Frankland (1825–1899) first expressed this idea in 1852:

> . . . the compounds of nitrogen phosphorus, antimony, and arsenic especially exhibit the tendency of these elements to form compounds containing 3 or 5 equivs. of other elements,* and it is in these proportions that their affinities are best satisfied; thus in the ternal groups we have NO_3, NH_3, NI_3, NS_3, . . . & c.; and in the five-atom group, NO_5, NH_4O, NH_4I, PO_4I & c. Without offering any hypothesis regarding the cause of this symmetrical grouping of atoms, it is sufficiently evident, from the examples just given, that such a tendency or law prevails, and that, no matter what the character of the uniting atoms may be, the combining power of the attracting element, if I may be allowed the term, is always satisfied by the same number of these atoms.

* Frankland begins by talking about equivalents which are empirical but slides almost imperceptibly into *atoms* (which are only hypothetical). His formulae are for equivalents.

Friedrich August Kekulé (1829–1896) spoke in terms of "affinity units" to express the number of equivalents with which elements and radicals combined. For example, he wrote, "The radical of sulphuric acid SO_2 contains three atoms, each of which is diatomic, thus representing two affinity units." Carbon, he thought, ". . . always combines with four atoms of a monatomic, or two atoms of a diatomic, element; that generally the sum of the chemical units of the elements which are bound to one atom of carbon is equal to four." He conceived of the double bond between two carbons, saying that two of the total of eight affinity units would be used to bond the two together, leaving six to be bound by other elements. "In other words one group of two atoms of carbon $= C_2$ will be hexatomic . . ." Each additional carbon would add two more affinity units to be bound, or n carbons would have $2n + 2$ unbound affinity units.

Kekulé and, independently, A. S. Couper (1831–1892) and A. Butlerov (1828–1886) also grasped the idea of carbon-to-carbon links in compounds containing more than one atomic weight of carbon—such links to use one of the four "affinity units" of each atom. Couper apparently wrote the first structural formulae representing links or affinity units or *bonds* by dotted lines. He represented ethyl alcohol (CH_3CH_2OH) as:

$$\begin{array}{l} \quad \ldots O \ldots OH \\ C^2 \\ : \ \ldots H^2 \\ : \\ C^2 \ldots H^3 \end{array}$$

using two oxygens each with weight 8 in place of a single oxygen of weight 16. This is the first of the structural formulae which were to appear. The modern representation is:

$$\begin{array}{ccccccc} & & H & & H & & \\ & & | & & | & & \\ H & - & C & = & C & - & OH \\ & & | & & | & & \\ & & H & & H & & \end{array}$$

The real development of the structural theory of organic chemistry came with Kekulé's discovery of the benzene (C_6H_6)

ring. There were six carbons to the molecule, which would require fourteen hydrogens ($2 \times 6 + 2$). According to Kekulé, in 1865 he was dozing in his study in Ghent dreaming of long chains of carbon atoms with their unsatisfied valences like so many snakes, when one "gripped its own tail and the picture whirled scornfully before my eyes." Representation of the six carbon atoms as a closed chain with alternate double bonds between alternate carbon atoms on the ring leaves six affinity units unsaturated. If the chain is open—that is, without a bond between the first and last carbons—there are eight unsaturated affinity units:

```
         H                          H
         |                          |
         C                          C
       /   \\                     //   \
  H—C         C—H            H—C         C—H
     ||        |       or       |         ||
  H—C         C—H            H—C         C—H
       \\   /                     \\   /
         C                          C
         |                          |
         H                          H
```

as benzene is represented today. Kekulé himself represented his closed chains at first by:

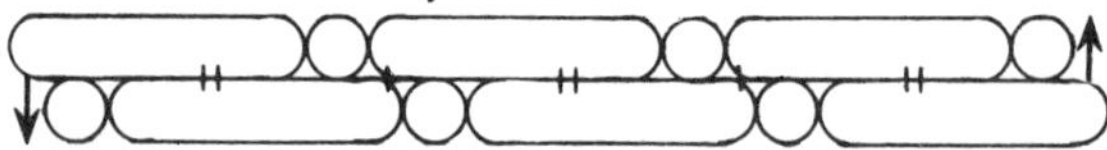

Kekulé had studied architecture and it was therefore no coincidence that his contributions to chemistry took a graphic and geometric form. He noted that with a single substitution of another element or group for a hydrogen all positions were equal, but with substitution for two hydrogens, the 12 and 2 o'clock could be replaced (ortho-) or the 12 and 4 o'clock (meta-) or the 12 and 6 (para-)—all others being equivalent to one of these three. Therefore substitution compounds with benzene for basis should exist in three *isomers,* three distinct chemical compounds with the same formula. Ortho-, meta-, and para-xylene is an example, with double substitution by CH_3 groups. Such a substituting group may also have a structure (side chain), and its hydrogens are also liable to replacement.

In 1874 Koerner identified the three dibromobenzenes by

counting the numbers of different tribromobenzenes which could be generated from each. The 1,4 DBr could generate only one TBr; the 1,3 could generate three different TBr; the 1,2 could generate only two.

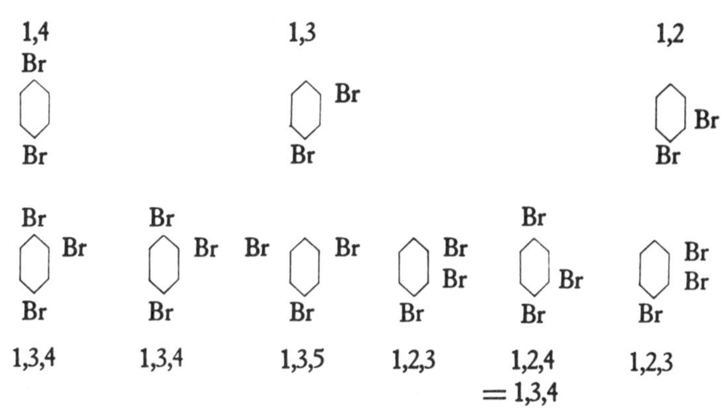

Erlenmeyer suggested in 1866 that naphthalene was a double ring– ; anthracene, known since its discovery by Dumas and Laurent in 1832, was shown to be a triple ring by Graebe and Liebermann in 1868. Heterocyclic compounds with atoms other than carbon in the ring were discovered in 1869. The *aromatic* compounds, so called because the familiar perfumes and flavors fit into this class, are characterized by closed rings. The fatty or *aliphatic* compounds are made of open chains of carbon atoms.

With the idea of molecular structure, the main outlines of organic chemistry were fixed. It was not enough to determine the composition of an organic compound or the exact proportions of its constituent elements or even radicals or types. This was little better than the fire, earth, air, and water of the alchemists. The molecular architecture, or structure, was the determining factor. The subsequent history of organic chemistry is the gradual mapping of the possibilities for arrangement of carbon compounds of increasing complexity. If we were to make a comparison, organic chemistry in 1869 would resemble architecture on the eve of the discovery of the brick. On the one hand there are the possibilities for construction offered by this brick, the carbon atom, with virtually no properties of its own but with limitless possibilities for towers, shelters, bridges, and so on. On the other hand there is the side of organic chemistry comparable to archeology. Although there is no limit to the number of possible arrangements of bricks or carbon atoms, an enormously complex architecture already exists, in the form of natural compounds waiting to be explored. In this sense biochemistry, the chemistry of life, forms a large, but at any given moment finite, subdivision of the infinite field of organic chemistry.

In 1874 an important pamphlet entitled *Chemistry in Space,* by the Dutch chemist J. H. van't Hoff (1852–1911), who had studied with Kekulé at Bonn, made the transformation from stoichiometric chemistry to stereochemistry complete. His work was based on an intriguing discovery by Pasteur. Louis Pasteur (1822–1895) had had difficulty in obtaining admission to the École Normale in Paris, his examiner noting that he was but mediocre in chemistry. "Will," Pasteur wrote later, "opens the door to brilliant and happy careers; work surmounts them."

Pasteur's first research dealt with the puzzle of the isomers of tartaric acid. These had been noted by Berzelius in defining isomerism in 1827. Two acids of the same composition may be obtained from the lees of wine—tartaric acid and racemic acid. The physicist J. B. Biot (1774–1862), in his studies of polarized (or directionally constrained) light, had observed that solutions of tartaric acid as well as various sugars and other organic compounds had the property of rotating the plane of polarization. Such substances Biot described as "optically active."

Ordinary crystals or their solutions may or may not restrict the vibrations of the light passing through them to a single plane as does the polaroid film in sunglasses. Optically active crystals or solutions rotate the plane of polarization of a beam of light. If the beam enters with the plane of polarization in a fixed direction, it leaves with the plane turned in a clockwise (dextrorotatory) or counterclockwise (laevorotatory) sense, depending on the crystal. The amount of rotation depends on the thickness of the crystal or solution traversed.

Eilhardt Mitscherlich (1794–1863), the rediscoverer of Boyle's old concept of mixed crystals (isomorphism),* noted in 1844 that while tartaric acid and its salts were all dextrorotatory, racemic acid and its salts were optically inactive. This was the problem which Pasteur undertook to solve. By slow crystallization of the sodium ammonium salt of racemic acid (racemate), he obtained well-formed crystals which he observed had modifying facets either to the right or to the left (1848). This, the phenomenon of hemihedralism (half-symmetry) or *enantiomorphism* (right- and left-handedness) of crystals had been observed by Haüy for quartz. The ordinary hexagonal symmetry of quartz is reduced by the presence of tiny modi-

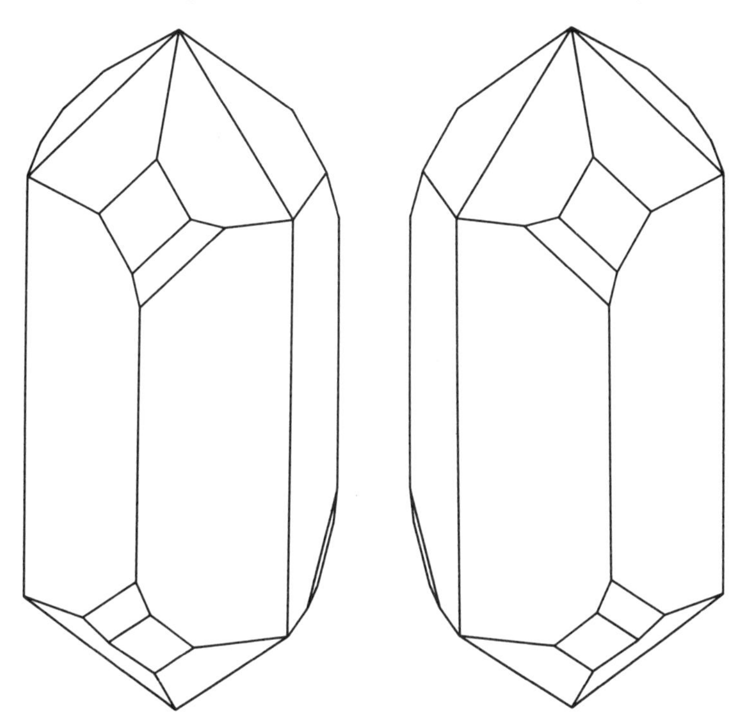

Figure 9. Left- and Right-handed Quartz—Haüy's Enantiomers From Bravais, 1866.

* Independently discovered by the mineralogist F. Beudant (1787–1850).

fying facets much like the hooked arms of the swastika. These give the crystals a screw sense, or polarity; like the right-handed and left-handed glove, the one is the inverse or mirror image of the other. No process other than reflection in a mirror can convert the right-handed screw into the left.

Biot had determined that right-handed crystals of quartz were dextrorotatory and left-handed ones were laevorotatory. Pasteur separated the right-handed crystals of racemate from the left and examined their solutions.

> I then saw with no less surprise than pleasure that the crystals hemihedral to the right deviated the plane of polarization to the right and that those hemihedral to the left deviated it to the left; and when I took an equal weight of the two kinds of crystals, the mixed solution was indifferent towards the light in consequence of the neutralization of the two equal and opposite individual deviations.

The right-handed enantiomer was identical with the tartrate. Pasteur had proven that the isomer, racemic acid and its salts, consisted of equal mixtures of right-handed and left-handed molecules, the right-handed occurring naturally as tartaric acid and its salts, the left-handed sugars being unknown in nature. When this was demonstrated to Biot, he broke down and exclaimed to Pasteur, "My dear boy, I have loved science so much in my life that this makes my heart pound."

In casting about for a chemical means of obtaining the laevorotatory (left-handed) enantiomer, Pasteur after great effort found that a penicillin mold would selectively destroy the dextrorotatory (right-handed) fraction. Until this time a completely mechanistic explanation had prevailed for the most fundamental processes of practical chemistry—the leavening of bread, brewing, the souring of milk and cream, cheesemaking, the fermentation of wine, as well as the analogous processes of putrefaction and decay. "The acetous fermentation is nothing more than the acidification, or oxygenation, of wine produced in the open air by means of the absorption of oxygen," Lavoisier had written. "The resulting acid is the acetous acid, commonly called Vinegar. . . ." The use of yeast or suitable ferment in these processes had been known since the time of the Egyptians but its function was believed to be only ". . . to give a commencement to the fermentation."

Pasteur changed all this. Yeast and related organisms perform the conversion from sugars and starches to alcohol and vinegar. The realization that these were biological processes and the dramatic demonstration of the idea was the work of Pasteur. The must of the grape would remain unchanged, milk would remain sweet indefinitely at any temperature, if it had first been sterilized and was kept sealed. Until this discovery of Pasteur's no food had ever been canned. A chemist, a crystallographer, he nevertheless undertook to apply his ideas to the epidemic which was then ruining the French silk growers. The diseases which he himself brought under control or cure after that of the silkworm include chicken cholera, anthrax, and finally hydrophobia (rabies). The application of his ideas to surgery by Lister made the modern hospital possible. In our time the use of molds as selective specifics against bacteria—particularly the use of Pasteur's penicillin—has been revived with spectacular success.

Pasteur's vitalism amounted to a subtle shift in the grounds of the debate, a shift which was hardly noticed by the participants but which made the arguments over the possibilities of synthesis essentially irrelevant. For it is no longer a question of whether fully inorganic, mechanical processes can produce alcohol from sugar and water, but rather of the extent of the role played by vital processes in the chemical world. For example, many of the vast beds of limestone which occur in the earth's crust are clearly and visibly accretions of lime obtained from sea water by animals for their shells. There is now a debate over the origins of the fine lime ooze and the beds of carbonate rocks apparently derived from ooze. Does the precipitation of lime from sea water require the intermediary of microscopic organisms? Bacteria are known which depend in their life processes upon surphur and not upon oxygen. Are the vast deposits of pure surphur associated with the salt domes of the world derived from compounds of saline deposits by the action of such bacteria? What is the origin of the world's petroleum? Some of these processes are so little understood that mechanistic concepts of hydrocarbon and carbohydrate formation in the upper atmosphere, or from meteoric invasion, have been proposed recently in works of great popularity and wide dissemination (for example, those of Velikovsky).

Pasteur was vitalist enough to write in 1860 that asymmetry was the prerogative of the organic kingdom and the distinction between living and dead matter. Just as all tartaric acid is dextrorotatory, or right-handed, this exclusion of one enantiomer or the other is characteristic of all natural organic compounds. All the natural amino acids, the principal complexes which make up the proteins, are left-handed.* Since the two kinds of molecules differ only in their right- or left-handedness, their energies must be the same and syntheses should in theory, and do in practice, yield racemates—equal mixtures of the two. Yet in our own chemistry we absorb only dextrorotatory glucose, and in fact the action of certain deadly poisons is believed

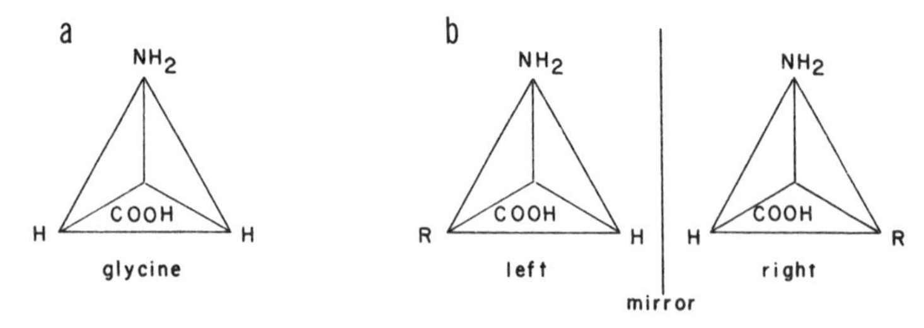

Figure 10. The Tetrahedral Carbon Atom
The carbon tetrahedron explains the right- and left-handed optical properties of the amino acids (optical isomerism). All the amino acids are based on the same central tetrahedral carbon atom, with hydrogen at one corner, an amine group (NH_2) at another, a carboxyl group (COOH) at another, and a residue (R) at the last. Two configurations are possible for all the amino acids except glycine (a), for which the residue is a simple hydrogen.
Turning one of the tetrahedra around or over will not bring it into congruence with the other because of the reversed positions of the residue (R). The two (b) are mirror images or enantiomers, like right- and left-handed gloves.

* Except glycine, which is symmetrical.

to be in their polarity, which like Pasteur's penicillin acts selectively on certain enantiomers.* The same amino acid, phenylalanine, which in the dextro form produces insanity, is harmless in the laevo form. Allergic reactions and responses to insect and snake venoms may be of similar nature.

It is difficult to escape the conclusion of Pascual Jordan that the origins of life lie in one or at the most a very few chance events. It may be that "only asymmetry can beget asymmetry," as Japp (quoted by Weyl in his *Symmetry*) believed, but it has been suggested that asymmetry in the environment, in circularly polarized ultraviolet light, for example, would produce asymmetric syntheses. According to Oparin and Fesenkov, "It has been demonstrated that this kind of light really existed under the natural conditions of the as yet lifeless earth and, hence, asymmetric amino acids could here too have originated abiogenetically." Asymmetric syntheses have actually been achieved by A. Terentyev and his associates, according to the same authors, by reactions performed on the surface of optically active crystals of quartz. The only trouble with this is that the earth's quartz is most certainly racemic and should not even locally exhibit asymmetry. More likely, the ancestral protoplasm was the ruthless survivor of a war to the death, pitilessly exterminating its mirror twin, the lost ancestor of a never-to-be world of laevo sugars and dextro amino acids.

A geometric explanation of this enantiomorphism was given by van't Hoff at the age of twenty-two in 1874, the year he completed the work for his doctor's degree. Van't Hoff and, independently, J. A. LeBel (1847–1930), who had been a student with van't Hoff, saw the connection between enantiomorphism and the fourfold (tetrahedral) model of the carbon atom which had been proposed by Kekulé (and also by Butlerov, as an outgrowth of his pioneering researches in chemical structure).

The four "affinities" of the carbon atom gave it an effective tetrahedral shape. Van't Hoff saw that

* Natural amino acids including those of bacterial proteins are laevo. The antibiotics contain dextro amino acids.

> ... When the four affinities of an atom of carbon are saturated by four different univalent groups, two and only two different tetrahedra can be obtained, one of which is the reflected image of the other....

To the possibilities of rearrangement of atoms with their affinities or bonds in the plane (structural isomers), there were now added the possibilities of arrangement in three-dimensional space and the possibilities of arrangement of complexes of atoms in space (stereoisomers). After the work of Couper, Butlerov, and Kekulé, it was possible to visualize a chain of carbon atoms with various bonds above and below the chain:

```
    H  H  H  H
    |  |  |  |
 H—C—C=C—C—H        (butene)
    |        |
    H        H
```

After van't Hoff and LeBel it became possible to visualize two stereoisomers of butene, *cis*-butene and *trans*-butene.

Models for optically active molecules were right- or left-handed screws. Quartz (SiO_2), for example, was found (1925) to be a structure of linked spiral screws of tetrahedra of silicon surrounded by oxygen. In 1951 Pauling and Corey proposed a spiral or helix as the model for extended chains of amino acids known as polypeptides. Most recently Watson and Crick at Cambridge have designed a double-helix model for nucleic acid molecules (DNA), believed to be the carriers of genetic information in the genes within the cell nucleus. (See Chapter 14.)

The most complex of all organic compounds, and therefore the most difficult to understand and to study, are the proteins. They are at the same time the most difficult to resolve and also the most interesting, since they represent the highest degree of complexity and organization of matter yet achieved by the processes of evolution, or indeed, by any processes conceivable. Our present level of understanding of the proteins and of the mechanism of DNA and RNA is of very recent origin, but awareness of the problem goes back to the eighteenth century. It was perhaps the greatest triumph of classical organic chemistry that during the second half of the nineteenth century a basic understanding of the chemistry and structure of these natural materials was reached.

On the eve of the overthrow of the phlogiston theory, Pierre-Joseph Macquer (1718–1784) published the *Dictionnaire de Chimie* (1776), the last definitive work of the old terminology of chemistry. In the *Dictionnaire* Macquer identified a class of organic substance including albumen, the globulin of blood, and the casein of milk as *albuminous.* These materials (the white of an egg, for example), which coagulate or "denature" on heating, are the *proteins.* The earliest isolation of a protein was apparently in 1747, when J. B. Beccari (1682–1766) obtained *gluten* from wheat flour. By 1840 Liebig was to assert that the proteins, along with the sugars and starches, and the fats and oils, were common to all living things.

Their extraordinary complexity (single protein molecules may contain hundreds of thousands of atoms) almost guaranteed that the elucidation of their chemistry would not be swift.

Just twelve years before Wöhler's and Liebig's benzoyl research of 1832, Henri Braconnot (1780–1855) at Nancy had obtained a sweet crystalline compound, *glycine,* from the digestion of albuminous gelatine in acid. This was the first of the amino acids to be obtained from the acid hydrolysis of a protein. Braconnot had already obtained sugar from the action of acid on wood, bark, straw, and hemp. Now he was trying to extract sugar from albuminous substances by the same means. He also succeeded in obtaining a white crystalline substance, which he called *leucine,* by the action of acid on muscle tissue.

After Braconnot's discovery of glycine there were further attempts to analyze proteins. In 1846 Liebig obtained *tyrosine* as a fusion product of casein and alkali. Horsford, in Liebig's laboratory (and independently Laurent and Mulder), found the formula $C_2H_5NO_2$ for glycine, which had been so named by Berzelius, according to the custom of giving the nitrogen compounds names ending in *ine.* G. J. Mulder (1802–1880) found both glycine and leucine in meat as well as gelatine and discovered the presence of nitrogen in addition to carbon, hydrogen, and oxygen. Mulder arrived at the formula $C_{40}H_{62}N_{10}O_{12}$ for what he considered (following the example of Wöhler and Liebig with benzoyl) the organic radical in albuminous substances.

> The organic substance which occurs in all parts of the animal body and as we shall soon see, in the vegetable kingdom as well, may be named *Protein* after πρωτος, *primarius*. Thus fibrin and egg albumin have the formula Pr + SP, serum albumin Pr + SP.

The P and S represent phosphorus and sulphur. Although Liebig soon disproved his formulae, Mulder's idea of the chemical unity of animal and vegetable protein and the fundamental importance of the proteins for living things was formally adopted by Liebig in his classification of the substances of living things into proteins, carbohydrates, and fats.

Justus von Liebig began life as a pharmacist's apprentice at Appenheim, near Darmstadt, Germany. He took his doctorate at Erlangen but, complaining there were no laboratories in Germany, made his way to Paris, where in Gay-Lussac's laboratory he did the analysis of fulminic acid which brought him into contact with Wöhler. Liebig has been described as hot-tempered and aggressive, but it was in his journal that J. R. Mayer's paper presenting the law of conservation of energy was published after it was rejected by Poggendorff's *Annals of Physics* (see Chapter 11).

Liebig's interest in chemical physiology and his researches establishing oxidation of foodstuffs within the organism as the source of animal heat disposed him favorably toward Mayer's work. His researches on plant nutrition were the foundation of modern scientific agriculture and in particular of the chemical fertilizer industry. Reasoning that the supply of minerals in the soil is necessarily limited, he undertook to supply by manures those minerals which analysis of plant ashes showed to have been removed. Unfortunately he provided the minerals in an insoluble form, to prevent their being washed away, and the sorry results attendant upon following his instructions were for many years a serious handicap to scientific agriculture.

In the early nineteenth century the subjection of the arts of chemistry to prolonged investigation produced the beginnings of their refinement into crafts. Pharmaceutical chemists sought to disentangle the active ingredients of drugs from the irrelevant complexities of their natural origin. Serturner extracted morphine from opium in 1805. A series of alkaloids were isolated. Dyes were isolated from their organic matrices.

M. E. Chevreul (1786–1889) showed that soap was formed by the combination of alkali with an acid constituent of fat, releasing glycerol. In 1832 he obtained a white crystalline nitrogen compound, *creatine,** from muscle tissue. Berzelius, unable to repeat his experiments, would not accept it as a true organic compound. The controversy, which also involved Wöhler, was settled when Liebig published "The Ingredients of Meat Extract" in 1847. He had found not only creatine but an organic acid which he called *inosinic acid* and which later proved to be the clue to the chemistry of the cell nucleus.

Auguste Cahours (1813–1891) first realized the structure of an amino acid when in 1857 he recognized in the glycine formula the acetic acid type.† The structural formulae for all the amino

---

* An American government chemist recently found that the only active component in the controversial cancer specific, Krebiozen, was a trace of common creatine. Krebiozen had been reported to be extracted from horse serum by a secret process.

†

```
       H                  H
       |     O—H          |     O—H
   H—C—C<            H—C—C<
       |     =O           |     =O
       H                 NH2
   acetic acid          glycine

       H
       |     O—H
   R—C—C<
       |     =O
      NH2
   amino acid
```

The COOH or carboxyl group is characteristic of all the organic acids. The other carbon, called the α-carbon, is linked to a radical plus a hydrogen plus the amino group NH_2. In glycine the radical is a simple hydrogen. Twenty amino acids have been obtained from the hydrolysis of natural proteins. All except two with an α-imino group (CHNH instead of $CHNH_2$) have the

acids conform to the same type. Each includes a head of acid (COOH) or *carboxyl* and an amine tail (NH_2). Since the amino acids are obtained from the proteins by the attack of acids, alkalis, and enzymes, early ideas of protein structure all assumed that the acids were the structural units of which the proteins were composed.

Attempts to reverse the denaturing process—the coagulation which characterizes the proteins on cooking—were not particularly rewarding until, in 1901–1902, Hofmeister and, independently, Emil Fischer (1852–1919) established the idea that the amine tail of one amino acid could link with the acid head of another, liberating a molecule of water and forming a CONH link which Fischer called the *peptide* link. A similar link was known in biuret, a double urea molecule giving a violet color indication called the biuret reaction. Proteins, peptides, and related compounds also gave the biuret reaction, and it was this clue which led Fischer to the successful linkage of two glycines. By 1907 he had joined fifteen glycines and three leucines in a peptide-chain molecule. Far short of a natural protein, these peptide syntheses nevertheless were the first manifestations corroborating the forebodings of the Luddites. Man, having learned to master Nature, would not stop now short of the creation of life itself.

Although these developments could not have been foreseen in 1874, the extraordinary successes attendant upon the chemistry of the period had given rise to a boundless optimism, and it is hard to believe that Pasteur or van't Hoff would have been surprised by the work of Watson and Crick. For Pasteur himself, nothing less than the full achievement of the Baconian prophecy of the New Atlantis would have sufficed. For this generation organic chemistry had reached the stage of mechanism foretold by Hooke. The properties of matter were to be explained by number and geometry:

> Those effects of bodies which have been commonly attributed to qualities, and those confessed to be occult, are performed by the

---

general formula above. All were synthesized before 1922. All occur naturally in the laevo form except glycine, which is symmetrical because of the hydrogen radical.

small machines of nature . . . the mere products of motion, figure, and magnitude.

The success of organic chemistry was not simply intellectual. Marcelin Berthelot (1827–1907), following Wöhler's synthesis of urea, had successfully synthesized ethyl alcohol and a series of synthetic fats, *synthetic* now in the sense of being compounds not known in nature. Before Berthelot, the natural arrangements of this brick, the carbon atom, had been studied in detail. Berthelot was the first chemist with the audacity to arrange the bricks for himself—to create. Almost in a spirit of piety, Berthelot undertook detailed studies of the history of chemistry and of alchemy in particular, as if he were repaying a debt of whose existence modern science is almost unaware.

The major practical achievement of organic chemistry in the nineteenth century came in 1856. William Perkin (1838–1907), an eighteen-year-old student trying to synthesize quinine, discovered aniline purple, the first synthetic dye and the beginning of the vast chemical empires first of Great Britain and, in emulation, of Germany. Legend has it that he made his discovery while trying to clean his glassware with alcohol. Perkin and others discovered synthetic dyestuffs and synthetic perfumes and flavors, and synthesized natural substances. The great commercial empires and consequent rivalries of the European powers were in large part based on the synthetic dyestuffs industries that grew out of Perkin's discovery.

With the development of stereochemistry the discovery and synthesis of useful materials passed from the realm of accident, of fine art, to the realm of science. Specific properties could be associated with specific molecular configurations. Rubber, for example, consists of C_5H_8 isoprene units (discovered in 1879 by Bouchardat) arranged in long coils, literally molecular springs. Chemists learned how to open rings and add chains; to join simpler molecules (monomers) into large molecules (polymers).

In these first practical achievements of organic chemistry, science in mid-nineteenth century justified the highest hopes of the idealists and the scholars, of those who tried by studying their environment to understand it, and by understanding, to master it. It was not by the abdication of their responsibility, by retreat into vitalistic expressions of unknowability, that the men of the nineteenth century developed organic chemistry. "It is necessary to work," Pasteur exclaimed on his deathbed, carried back perhaps to the terrible defeat of France in 1870 and his resolution to pay off single-handedly the crushing indemnity that Germany imposed.

The reforms introduced into French agriculture, brewing, distilling, and preservation of foodstuffs by his new biochemistry easily paid the indemnity, but that achievement is dwarfed by the contribution that his germ theory of disease has made to human well-being.

9

Origins of Electrochemistry

HIS WIFE TO WHOM HE WAS MOST TENDERLY ATTACHED, BEING IN A DECLINING STATE OF HEALTH, USED A SOUP MADE FROM FROGS AS A RESTORATIVE; AND SOME OF THESE ANIMALS, SKINNED FOR THE PURPOSE, HAPPENING TO LIE ON A TABLE IN GALVANI'S LABORATORY, ON WHICH WAS PLACED AN ELECTRICAL MACHINE, ONE OF THE ASSISTANTS IN HIS EXPERIMENTS, BY ACCIDENT, BROUGHT THE POINT OF A SCALPEL NEAR THE CRURAL NERVES OF A FROG LYING NOT FAR FROM THE CONDUCTOR.

—CHALMERS, 1812

Few single trifles in history have had such momentous consequences or set into motion such a chain of events as the twitching of a frog's leg in the anatomy laboratory of the University of Bologna, in Italy, late in the eighteenth century. The electrostatic machine had been used at this time to generate powerful static charges on the plates of Leyden jars, and Franklin had established that lightning was the discharge of just such an electrical accumulation. The electric shock had been observed in the discovery of the Leyden jar (1745), and its therapeutic, moral, and comic aspects exhaustively explored in the ensuing years. In 1773 Walsh and Ingenhousz* were

* Jan Ingenhousz (1730–1799), Dutch physician and chemist, was the first to appreciate the power of plants to absorb CO_2 and give off oxygen. John Walsh (d. 1795), an English M.P., published a paper "On the Electric Property of the Torpedo" in the *Philosophical Transactions of the Royal Society* of London in 1773.

demonstrating that the numbing attack of the torpedo-fish was an electric shock. Luigi Galvani (1737–1798), physicist, *accoucheur,* and professor of anatomy at Bologna, was investigating the effects of electricity on the animal organism.

Galvani's initial discovery was, he tells us, an accident. He had dissected a frog and one of his assistants had provoked a violent convulsion of the leg muscles by a chance touch of a scalpel.

> Another of those who used to help me in electrical experiments thought he had noticed that at this instant a spark was drawn from the conductor of the machine. . . . I myself was at the time occupied with a totally different matter; but when he drew my attention to this, I greatly desired to try it for myself, and discover its hidden principle. So I, too, touched one or other of the crural nerves with the point of the scalpel, at the same time that one of those present drew a spark; and the same phenomenon was repeated exactly as before.

Galvani found that the frog preparation served equally well as a detector of lightning discharges, and then tried to detect atmospheric electricity in calm weather, since he had observed "that frogs which had been suitably prepared for these experiments and fastened by brass hooks in the spinal marrow, to the iron lattice round a certain hanging garden at my house, exhibited convulsions not only during thunderstorms, but sometimes even when the sky was quite serene."

Bit by bit in the decade which followed, Galvani varied his experiments until he could finally demonstrate that connecting the nerves to the muscles with an external conductor, especially one made from two dissimilar metals, would produce the convulsion. His view was that a fluid animal electricity was passed through the metal. This idea was attacked immediately by Alessandro Volta (1745–1827), who accounted for the phenomenon entirely by the action of the metals. Volta was able to generate what he thought of as a circulation of the electric fluid (current) in a circle of different conductors, some of which (the electrolytes) were moist.

"Do not ask in what manner," wrote Volta. "It is enough that it is a principle and a general principle."

In his eagerness to attack all traces of vitalism in this phenomenon (which was called Galvanism after its discoverer), Volta missed its essential chemical nature, even after this was pointed out to him in 1796. Northern Italy was overrun by French armies. ". . . the Directory in Paris, giving itself the airs of a sovereign firmly enthroned, began to show a mortal hatred of everything that was not commonplace," wrote Stendhal. Galvani, who would not change his loyalties to fit the times, was stripped of his posts. Unable to bear poverty and disgrace, he died soon afterward. Volta was honored by Napoleon and promoted at the Austrian restoration.

"I dress myself up like an actor in a farce to win a great social position and a few thousand francs a year," Conte Mosca says in *The Charterhouse of Parma.*

The violence of the frog's convulsions, the galvanic action, as it was called, was the measure of the intensity of the effect. Volta's circles of conductors produced extremely weak effects but were continuous sources of current, as opposed to Leyden jars or static machines which produced strong effects lasting only an instant. In 1800 Volta hit upon the pile, or battery, as a means of multiplying the intensity of the electricity. Couples of different coins (copper-silver or copper-zinc) separated by moistened pasteboard could be piled up in any quantity. Connecting the top and bottom coins produced a shock, an effect which could be repeated indefinitely.

In March of 1800 Volta described his pile in a letter to Sir Joseph Banks, then president of the Royal Society, who immediately told William Nicholson (1753–1815). Repeating the experiment, Nicholson and a surgeon friend, Anthony Carlisle, used water at the upper plate of the voltaic pile to make a better contact and observed some effervescence. They repeated their experiments with care, collecting the gases that came off. By May 2 of the same year they had discovered the electrolytic dissociation of water into free hydrogen and oxygen. Beccaria, who had been Galvani's master, and Priestley too had observed electrolytic dissociation (as it was later called) using static electricity excited by friction, but in the absence of a continuous current the effect had been limited. Now William Cruickshank

(1745–1800) used the voltaic pile to decompose metallic salts, and Humphry Davy grasped the essential idea of the electrical nature of the chemical affinities.

First Davy demonstrated the chemical nature of galvanic action by showing that Volta's copper-zinc pile depended on the oxidation or corrosion of the zinc. The effect did not last forever, but only as long as the metallic zinc corroded.* To increase the effect by increasing the rate of chemical attack he tried strong acids and thus developed strengths of a new order of magnitude. He began the study of corrosion and was the first to hit on the idea of protecting a metal such as iron by making it the negative side of a couple, that is, by putting it in contact with a base metal such as zinc, which then corrodes in place of the iron.

Humphry Davy was born at Penzance, Cornwall, in 1778. After the death of his father, a woodcutter, young Davy was apprenticed to a pharmacist. Largely self-educated, he was fortunate enough to come in contact with a number of generous souls, including Gregory Watt, the son of James Watt, who taught him the chemistry of the times and found him a laboratory assistantship with Dr. Beddoes at Bristol. There he investigated the physiological effects of nitrous oxide (laughing gas) by breathing it himself. The gas was administered to volunteers in the public lectures at Count Rumford's new Royal Institution, according to a lady diarist quoted by S. C. Brown:

> The other day they tried the effect of the gas so poetically described by Beddoes; it exhilarates the spirits, distends the vessels, and in short, gives life to the whole machine. The first subject was a corpulent middle-aged gentleman who, after inhaling a sufficient dose, was requested to describe to the company his sensation. "Why, I only feel stupid." This intelligence was received amid a burst of applause, most probably not for the novelty of the information.

* In every voltaic cell the metal atoms of the positive side lose electrons (negative charges) and enter into solution or react as positive ions. Electrons enter the circuit, where they flow to the negative terminal (cathode).

Davy had also performed a crude experiment of melting two pieces of ice by rubbing them together, and on the strength of this he published a lengthy paper attacking Lavoisier's idea of the heat substance, caloric. Davy proposed that light, rather than caloric as Lavoisier had suggested, was the substance compounded with a base to form oxygen. Rumford, having himself used frictional heat as his argument against caloric, was certain that in Davy he had a scientific genius. He fired his lecturer, a Dr. Garnett, and appointed the twenty-three-year-old Humphry Davy in his place. It was 1801. Davy had already been drawn to the new phenomenon of electrochemistry and had in fact published his discovery of the increased power of batteries with acid electrolytes. This publication alone withstood the ridicule which greeted his early speculations on material light, into which he was led by his enthusiasm for Lavoisier and Rumford. They were his only venture into theory. He was never again to permit himself an enthusiasm.

Lavoisier's philosophy of science may have been positivist but his chemistry was highly speculative. The conservation of mass was the first principle of Lavoisier's philosophy, but the first principle of his chemistry was his stubbornly held conjecture that oxygen was the acid-former. From this he concluded that nitrogen, which he and the phlogistians agreed was opposite in its properties to oxygen, must be the effective ingredient in strong bases (such as lye, KOH). This was incorrect. At the same time he thought that the alkaline earths such as lime (CaO) and magnesia (MgO), being weaker, would consist of oxygen neutralized by being combined with a metal. This turned out to be correct. With his success in developing powerful acid electrolyte batteries and his realization that the chemical bond was essentially electrical, Davy undertook to test Lavoisier's idea of the strong bases, trying to decompose them electrically.

"The different bodies naturally possessed of chemical affinities appear incapable of combining, or remaining in combination, when placed in a state of electric difference from their natural order," he said in the Bakerian Lecture to the Royal Society for 1806. For this lecture Napoleon awarded him a three-thousand-franc prize, although England and France were still at war.

After repeated trials, Davy learned to decompose potash and soda by passing a strong current through moistened lumps of paste, using platinum for the contacts.

> The potash began to fuse at both its points of electrization. There was a violent effervescence at the upper [positive] surface; at the lower or negative surface, there was no liberation of elastic fluid; but small globules having a high metallic lustre, and being precisely similar in visible characters to quicksilver, appeared.

The complete account of the discovery and properties of the two new elements sodium and potassium made up Davy's Bakerian Lecture to the Royal Society in 1807.

By 1808 he had learned to decompose the alkaline earths to obtain magnesium, calcium, strontium, and barium, at first by electrolysis in an atmosphere of naphtha to take up the evolving oxygen, then by the electrolysis of a mixture of the earth and mercuric oxide which yielded an amalgam (solution of metal in mercury). The metals were obtained in pure form by heating the amalgam to distill off the mercury. Davy also thought he had obtained a metal *ammonium* when Berzelius and Pontin wrote from Stockholm that similar procedures yielded an amalgam from ammonia.

Davy was never able to reduce alumina, zirconia, or beryllia. Aluminum was not made electrolytically until the Hall-Heroult process was developed in 1886, but it was obtained by H. C. Oersted in 1824, when he reduced aluminum chloride with metallic sodium. Zirconium was obtained by Berzelius in 1825; he also obtained titanium (1824), silicon (1810), selenium (1817), and thorium (1828). Uranium (1789), tellurium, and chromium (1798) had been discovered by M. H. Klaproth (1743–1817). Davy himself isolated lithium and boron. Beryllium was not obtained until Wöhler isolated it in 1828.

In part Davy's results were due to his success in increasing the power of the batteries he used. In the Bakerian Lecture for 1806, in which he first proposed that the chemical bond was electric, he pointed out that the resolution of any body into its component elements must be a matter of a powerful enough battery and that the steady increase in technological power assured the eventual reduction of all compounds.

He had learned caution since the days of his unfortunate hypothesis of oxygen as a compound containing light, and retreated to empirical definitions. Elements were simply bodies which chemists had not yet learned to decompose. "There is no reason to suppose that any real indestructible principle has been yet discovered . . . " he wrote, and assigned completely neutral names to the newly discovered metals potassium, sodium, etc., signifying only that these bodies were derived from potash and soda. This caution and his developing powers as an experimental chemist led Davy in the years from 1809 to 1818 to the overthrow of what Lavoisier regarded as the cardinal point of his chemistry—the theory of acids and oxygen, that most unfortunately misnamed substance which from being dephlogisticated air had become the acid-former, while it was in fact neither the one nor the other.

Muriatic acid (our hydrochloric, HCl), or spirit of sea salt, had been an industrial commodity available in increasingly pure form since ancient times. It was originally obtained by distilling salt in acids such as vinegar. In Lavoisier's treatise on chemistry he lists the muriatic, fluoric, and boracic radicals along with sulphur, phosphorus, and carbon, as nonmetallic "simples" which join with oxygen to form acids. But Scheele, the phlogistonite who discovered oxygen, did not find it in muriatic acid.

In 1774 Scheele followed Priestley's method for obtaining pure HCl by the action of sulphuric acid on sea salt. He used this HCl on *black magnesia* (manganese dioxide), which was considered related to *magnesia alba.* He obtained a new metal—manganese—and a new gas—chlorine. The same reasoning that led Joseph Black to conclude that an air (CO_2) was driven by hydrochloric acid from a fixed state in magnesia alba led Cavendish to conclude that phlogiston (hydrogen) was driven from metallic zinc by the action of the acid, leaving a calx. Scheele, following the same phlogiston reasoning (calx + phlogiston = metal), thought of the phlogiston (hydrogen) as separated from the muriatic acid by reaction with the black magnesia to make the new metal manganese. In other words, the black magnesia drove off a new gas (chlorine) from marine acid. The black magnesia was the calx which took up phlogiston from the acid to make the new metal.

A better demonstration of the power of the phlogiston concept or the weakness of the oxygen theory in its eighteenth-century form could hardly be devised. Lavoisier, Fourcroy, and Berthollet proceeded to explain away Scheele's results, correctly interpreting the appearance of metallic manganese as a reduction from the oxide but erroneously concluding that the evolved oxygen combined with the muriatic acid to give a higher oxide of the muriatic radical, *oxymuriatic acid* (chlorine).

This was the background with which Davy, who had already successfully isolated the boracic acid radical (boron, 1807), began. Believing that muriatic acid was simpler than the greenish oxymuriatic acid gas (chlorine) which persistently turned up in his electrolytic experiments, he concentrated on the muriatic acid. Gay-Lussac and Thénard were trying at about the same time to reduce the oxymuriatic acid (chlorine), which, since it supposedly contained even more oxygen than muriatic acid, should have readily yielded up its oxygen to heated charcoal. They were the prisoners of their terminology, since they knew their failure to break down the green gas could have been explained only if it were an element.

Davy's work on electrochemistry had familiarized him with the notion of one metal replacing another in solution—as when an iron nail is dipped in a solution of a copper salt, metallic copper appears as a coating on the nail, its place in the solution taken by iron. In the same way a copper coin in a silver salt solution is soon coated with silver. Davy could not burn charcoal brought to a white heat in dry muriatic acid gas and therefore concluded that there was no oxygen in the gas. Seeing the evolution of free hydrogen through the action of the acid gas on metals, he reasoned correctly that the hydrogen was displaced from the acid gas by the metal just as silver is precipitated from solution by copper. With this and similar experiments the role of hydrogen rather than oxygen as the acid-former, or acidifying principle, was established.

Next Davy attacked the oxymuriatic acid gas which was evolved by the separation of hydrogen from muriatic acid. When this gas in turn would not yield oxygen to white-hot charcoal, or separate its mysterious base in electrolysis, Davy made the shift in thinking to which his researches all led him. The gas had to be an element. He proposed to call it *chlorine* for its color. "Few substances," he wrote, ". . . have less claim to be considered as acid than oxymuriatic acid . . . it is evident . . . that Scheele's view . . . of the nature of the oxymuriatic and muriatic acids, may be considered as an expression of facts."

Although Davy was knighted in 1812, elected to the presidency of the Royal Society in 1820, honored and enriched, successful in all he undertook, he was in his later years somewhat troubled by the rise of his protégé, Michael Faraday (1791–1867), the young man whom thoughtless persons described to Davy as his greatest discovery. Faraday's life, like Davy's, began humbly. His father was a blacksmith and raised Faraday in an atmosphere of hard work, self-denial, and religious nonconformity. A bookbinder's apprentice, Faraday so illustrated the industrious precepts of the fundamentalist sect (Sandemanians) to which he adhered that he read the books he bound. These led him to attempt experiments and to meet with scientific amateurs, one of whom, a Mr. Dance, took Faraday to Davy's lectures at the Royal Institution and encouraged him to apply to Davy with his lecture notes, a procedure well calculated to win Davy's approval. After several false starts Faraday became Davy's assistant in 1813, and shortly afterward, when Davy was married, Faraday accompanied the couple on a tour of Europe. Here Sir Humphry's relationship with Faraday entered an ambiguous stage. On the one hand he deeply offended his student by trying to use him as a valet. Yet on the same trip, perhaps in the sense of amends, he encouraged Faraday to undertake his first independent research.

Even before joining Davy at the Royal Institution, Faraday

had experimented with electrolysis, decomposing magnesium sulphate with a kitchen-made voltaic pile. Under Davy's direction he learned the experimental chemistry of the time. Davy, whose imagination was chemically oriented, looked for the differences in kinds of matter. Faraday, as he became more and more independent of Davy, developed the physicist's turn of mind.

After Davy's death, when Faraday undertook the serious study of electrolysis, it was the regularities, the identities in the process, on which he concentrated. Davy (and Grothus*) had referred to the electrolytic plates as *poles* and considered that they exerted electrical attractive and repulsive powers on the solution in the Nicholson cell in the same way that they would have attracted or repelled statically charged pith balls or bits of fluff in the air. But static electrical forces, like magnetic and gravitational forces, diminish with increasing distance, and Faraday showed that the rate of decomposition was not affected by the size or positioning of the plates, but by the electrical input alone. "Is the law this," he asked in his diary, "that whatever the size of plates or number intervening, . . . equal currents of electricity measured by the galvanometer evolve equal volumes of gas or effect equal chemical action in a constant medium?"

By 1832 Faraday understood that the products of decomposition were in quantities proportional to their chemical equivalence. "Think it will be very important to have a new relation of bodies under the term 'electro-chemical equivalents' tabulated," he entered in his diary. With William Whewell (1794–1866) he introduced the modern terminology. He changed the *poles* which implied attraction and repulsion to "electrodes," calling the positive pole the *anode* and the negative pole the *cathode*. He speculated on the mechanics of the decomposition, which he thought must be due to "an internal corpuscular action," and to these particle products of decomposition in the solution, the "true acting evolving bodies," he gave the term *ions;* the ions moving toward the anode he called *anions* and

* Theodore Grothus (1785–1822) published a memoir on electrolytic dissociation in the *Annales de Chimie* for 1806.

those moving toward the cathode were *cations*. The conducting solution was the *electrolyte* and the whole process he termed *electrolysis*. What he had established was essentially that the passage of a unit quantity of electricity (96,500 coulombs of charge), now called a *faraday,* would release at opposite electrodes equivalent weights of substances. The same amount of electricity would release eight grams of oxygen at the anode of an electrolytic cell and one gram of hydrogen at the cathode. In the decomposition of HCl it would be one gram of hydrogen and thirty-five grams of chlorine.

It was at this stage in his career that Faraday began to question the basic assumptions of mechanical materialism, specifically the concept of atoms and the void.

The physicist Helmholtz later pointed out that the faraday of electricity was a kind of gram-equivalent quantity itself particular or atomic, with an Avogadro's number of electrical particles (electrons). If each faraday were able to move an Avogadro's number of atoms of hydrogen, or the twenty-three-times heavier sodium, or the same number of atoms of potassium, toward the cathode, the faraday appeared to consist of that same number of electrons, each like a ticket on a bus, good for unit transportation regardless of the size, or the age, or the sex of the passenger. To Faraday, imbued by Davy with a distrust of hypothesis, this mechanical conclusion was highly dubious.

> . . . there can be no doubt, that the words definite proportions, equivalents, primes, etc., which did and do express fully all the *facts* of what is called the atomic theory in chemistry, were dismissed because they were not expressive enough . . . they did not express the hypothesis as well as the fact.

By 1833, when Faraday was publishing his law of electrolysis, he was already perhaps the most distinguished scientist in England, his fame resting on the dramatic discovery of the electric motor effect and on the even more remarkable discovery, in 1831, of the induction of electric current. The latter began, Faraday wrote, in the attempt to find a parallel to the phenomenon of induction as it was known for static charges.

If an uncharged body is brought near a static electric charge, an opposite charge is induced on its surface. Faraday thought that a current should be induced in a conductor brought near a wire carrying a current of electricity. On the day he began his experimentation he determined that a current was indeed induced momentarily in the conductor at the moment of turning the current in the primary circuit on or off. He accounted for this by describing the distribution of magnetic force about the current-carrying wire and showing quantitatively that the induction in a secondary circuit depended upon the relative motion of this magnetic force and the secondary circuit. The distribution of forces in space he described as the *field,* and he represented this by lines of magnetic force—closed curves tangent at every point to the force vector at that point.

In the original Newtonian or mechanical physics all phenomena were to be accounted for by the hypothesis of atoms or corpuscles, their motions, their number, and their configurations in the aggregate. Electricity was a fluid of these particles, subtle enough to flow through the interstices between the atoms of conductors. Heat had originally been the caloric fluid, magnetism also a fluid phenomenon. Newton, by the ingenious design of hypothetical particles, was even able to account for the most complex behavior of light. True, gravitation in the Newtonian mechanics appeared as a force between particles, acting instantaneously through empty space, but Newton was careful to explain in the *Principia* that these were empirical and mathematical laws neither expressing nor precluding hypotheses.

> . . . I frame no hypotheses; for whatever is not deduced from the phenomena is to be called an hypothesis; and hypotheses, whether metaphysical or physical, whether of occult qualities or mechanical, have no place in experimental philosophy . . . And to us it is enough that gravity does really exist, and acts according to the laws which we have explained, and abundantly serves to account for all the motions of the celestial bodies, and of our sea.

As parallels to the law of gravitational force between two bodies, with magnitude given by the product of the masses of the bodies divided by the square of the distance between their centers, C. A. Coulomb (1736–1806) had established similar laws for the forces of electric and magnetic attraction. When all phenomena were referred back to these unchangeable attractive and repulsive forces between bodies, with the intensities of these forces depending only on distance, the task of physical science would be complete, according to Helmholtz.

In 1820 H. C. Oersted (1777–1851) first recorded observations of the magnetic force accompanying a current of electricity. It was immediately clear that these forces, at any rate, were geometrically distinct from the inverse square forces which had been described by Newton and Coulomb. A magnetic needle placed next to a current-carrying wire is neither attracted nor repelled by the wire, but instead turned in a direction bearing the same relationship to the wire as the thread bears to the axis of a screw. In Faraday's induction of a current of electricity, the velocities of the moving parts were critical. The simple relationships of inverse square laws had nothing to say about these phenomena, and in fact the law of electrolysis which Helmholtz thought demonstrated the corpuscular character of electricity had been found as a result of the failure of the inverse square law to account for electrolysis. Not the distances between electrodes or the sizes of the electrodes, but the quantity of current passed, was the determining factor.

Faraday still used the terminology of particles and atoms in the 1830's, but his researches on conductivity could not be explained in these terms. His particles, he thought at one time, were thrown into an "electrotonic" state. "The induction is carried through them [insulators] from particle to particle, and not at once by a single act . . ."

Instead of a fluid of material particles, electricity in Faraday's thinking became a state of tension, its passage a flow of force. The compass needle turned because the current in the wire produced an elastic strain in the intervening Aristotelian space. Connecting a battery to an electrolytic cell put the matter of the electrolyte into a state of tension analogous to a stretched spring, which Faraday referred to as a state of polarization. "The atoms of matter are in some way endowed or associated

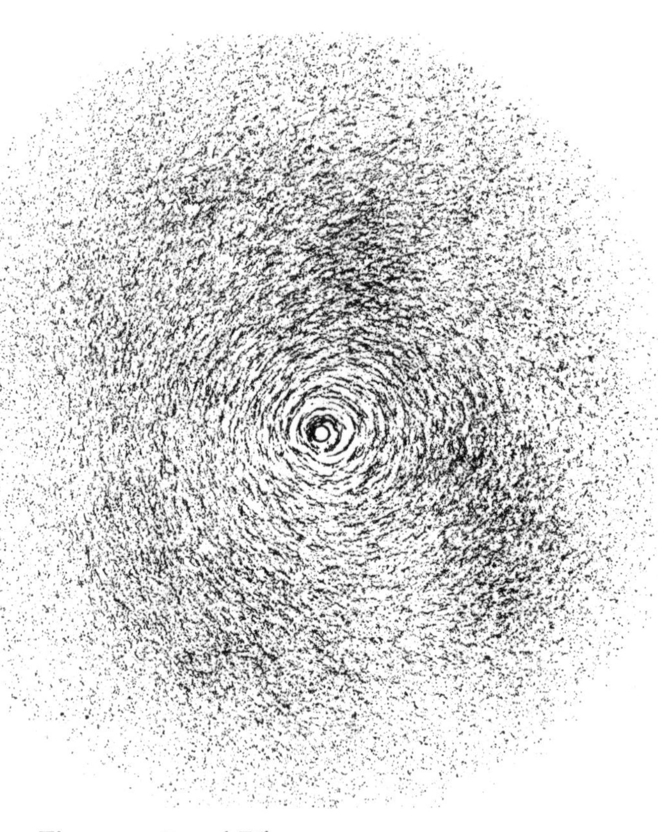

Figure 11. Oersted Effect
A wire through which an electric current is flowing comes up through the center of the diagram. Iron filings line up tangent to circles concentric about the wire, tracing the grain of the magnetic field.

with electrical powers," he wrote, "to which they owe . . . their chemical affinity." This was why the decomposition of a definite quantity of electrolyte corresponded to a definite quantity of electrical flow. It was the same quantity which would be generated in a voltaic cell by the formation of the same quantity of the compound. "But I must confess I am jealous of the term atom," he added. In place of material atoms and forces acting between them, Faraday inclined toward the Leibnizian monads of pure force, as these were constituted in the works of the Jesuit physicist-philosopher Boscovich.

Ruggiero, or Roger Boscovich (1711–1787), attacked the foundations of atomism and materialism, principally on the illogicality of action at a distance. In place of Democritean atoms—small, impenetrable, permanent bits of matter—he proposed a universe of point centers of force, these points having the property of inertia.

> The forces are repulsive at very small distances, & become indefinitely greater and greater as the distances are diminished indefinitely . . . When the distance between them is increased, they are diminished in such a way that at a certain distance which is extremely small, the force becomes nothing. Then as the distance is still further increased the forces are changed to attractive forces, . . . when we get to comparatively great distances, they begin to be continually attractive & approximately inversely proportional to the squares of the distances.

The whole of the universe was continuous and full. There were not two entities, atomic matter and void, with the motion of the atoms accounting for forces. The universe was a continuous field of force, thickening and thinning. This force-stuff was concentrated in points which by their aggregation accounted for all the gross observable properties of matter. As a large number of Boscovichian points came together their forces were added together so that collectively they presented the attributes of weight, hardness, and impenetrability.

This concept provided a ready answer for Faraday's difficulties with electrical and magnetic fields. Space itself was the elastic medium for the transmission of action. If the chemistry of the eighteenth century could regard solid, material magnesia alba as three-fourths made up of air somehow condensed and "fixed" in the rock, the 1758 theory of Boscovich required only the concept that the "air," as well, was a concentration of pure forces. In place of many different kinds of atoms and molecules, the Boscovichian doctrine required only pure force; the heterogeneity of the world was to be accounted for by num-

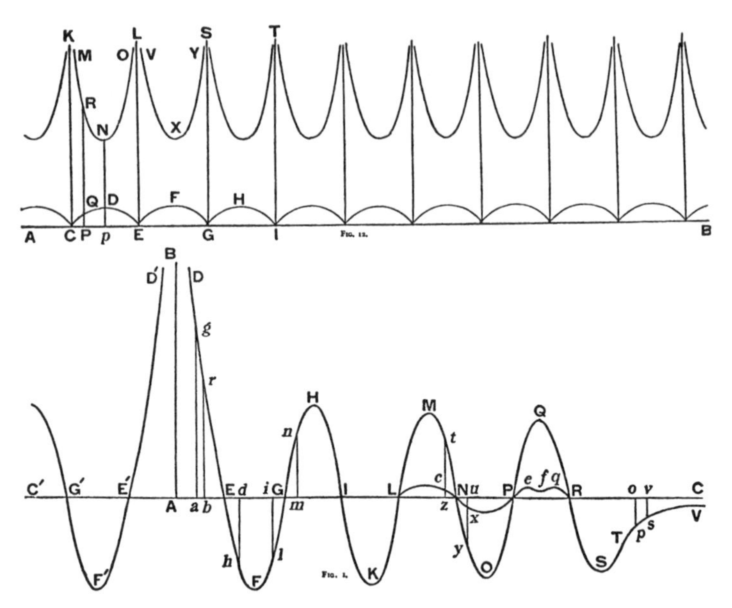

Figure 12. Boscovich's Point Centers of Force, 1763
Instead of material atoms and forces between them, Boscovich's world is made solely of force centered on points in space. At each center the force is repulsive, then attractive, then repulsive, etc., and finally an inverse square attractive force. The upper curve illustrates how the point centers could be positioned where attractive and repulsive forces balance at short distances apart to make extended matter obeying Newton's inverse square law of gravitation. Diagrams like these now appear in textbooks to illustrate the modern theory of the solid state.

ber and geometric form superposed on the elemental natural principle. As a screw forced through a pine plank sets up strains in the wood beyond its immediate path, and not parallel to its direction of advance, Faraday visualized the electric current as a stream of point centers of force setting up spreading tensions in the surrounding elastic media.

Boscovich's concept, seized upon by Faraday and elaborated into the concept of the field, was to provide the basis for the major developments in physics of modern times. We need only mention Maxwell's equations of the field and the discovery of electromagnetic radiation by Hertz, the theory of relativity, and the present development of quantum field theory, to justify the significance which should be attached to the *archei* of the Basler Bombastus and his successors. Nevertheless, in chemistry and the material sciences the philosophical advantages inherent in the abandonment of the illogicalities of atomism—the action at a distance, the indivisible point with breadth and parts and shape, the plurality of fundamentals—were subordinated to the nascent positivism of the age, rejecting all hypotheses atomic and continuous, mistrustful of speculation, and preferring to concentrate on the empirical and the practical.

The theory of solutions was not completed by the work of Faraday. It continued to attract the attention of the ablest scientists of the century. As Faraday visualized it the decomposition of a salt in solution was accomplished by the electric current, which was also required to move the ions to opposite electrodes. Yet such a motion of ions (charged particles) was itself a current and accounted exactly for the current of the electrolysis. The obvious but implausible conclusion that the dissolving of the salt was itself a decomposition into ions, was drawn by Rudolf Clausius of Zurich, among others, and was finally proved in 1887.

In a number of ways the same kind of ideal behavior could be observed for solutions as for gases. As in Dalton's law of partial pressures (see page 109) the pressure of a mixture of gases was determined by the equivalent fractions of each gas in the mixture, so the solutes (dissolved substance) affected the properties of the solvent (dissolving substance) according to their equivalent fractions. For example, the addition of one equivalent weight (gram molecular weight or *mole*) of alcohol to a large volume of water lowers the freezing point to the same degree as the addition of one mole equivalent weight of sugar to the same volume of water or one mole equivalent weight of any solute. This lowering of freezing and boiling points is due to the lowering of the pressure of the vapor of the solvent (Raoult's Law). In the same way osmotic pressure, the pressure of mixture of solutions of different concentrations through a semipermeable membrane, is also determined by the mole fraction—or in Avogadro's terms, the number of molecules—independently of their kind or chemistry.

In contrast to this molar behavior of dilute solutions of organic solutes, van't Hoff found that acids, bases, and salts (electrolytics) behaved in solution according to the number of moles of their constituent radicals. One mole of NaCl, for example, in dilute solution lowers the freezing point as much as two moles of alcohol; one mole of $NaHSO_4$, as much as three. By relating these measurements quantitatively to measurements of the conductivity of electrolytes, Svante Arrhenius (1859–1927) established the modern theory of electrolytic dissociation, according to which an acid, base, or salt in dilute solution spontaneously dissociates into ions. Difficult as it was to accept the idea that molecules as such (NaCl, K_2SO_4) do not exist as entities in the solution but rather as ions (Na^+, Cl^-, K^+, SO_4^{--}), a number of independent observations by Wilhelm Ostwald confirmed the idea; and it is now known that even in the solid, it is as ions rather than molecules or even atoms that the components of strong electrolytes exist. Detailed calculations by Debye and Hückel in 1923 accounted for the departures from ideal dissociation by the interactions between oppositely charged ions.

With the work of Faraday and the long line of investigators, not a tenth of whom we have considered, on the theory of solutions, we glimpse, although still strictly within the bounds of classical chemistry, a totally different kind of science, one which by the middle of the nineteenth century was beginning to impress itself upon the physicists as it previously had upon the mathematicians.

The earliest chemical ideas had concerned gross problems and had been settled by what Francis Bacon referred to as a *crucial* experiment. When Lavoisier used the burning glass on tin in a sealed vessel, then opened the vessel so that air rushed in to take the place of the oxygen which had combined with the tin, to show that the total gain in weight equaled the gain in weight of the calcined tin, he proved conclusively that the tin in calcination had gained matter from the air rather than losing phlogiston to it. But in the establishment of the dissociation theory we see that freezing-point measurements, conductivity, osmotic pressures vary in a complex and not obvious way with the saturation of the solution and the chemistry of the electrolyte.

Assuming a model of ions in solution attracting an atmosphere of oppositely charged ions and considering the behavior of this assemblage in an electric field, Debye and Hückel emerge with a complicated algebraic expression predicting the variation of conductivity with concentration. When the predictions of this theory are compared with experimental measurements, the close agreement is a measure of the "goodness" of the theory. To speak of the truth of the theory is to lapse into metaphysics. The original equations of Arrhenius also show agreement with experiment, but the agreement is better for weak electrolytes than for strong electrolytes, and in any case it is not so high as that of the Debye-Hückel theory.

Does this prove, then, a detailed picture of positive ions struggling to break through a deformed cloud of negative ions to reach the anode? It would be rash indeed to accept this or any detailed model as ultimate reality or as anything other than a convenient translation into familiar pictorial terms of a set of mathematical relationships between various observable properties in our experiments.

"We no other way know the extension of bodies than by our senses," Newton wrote in 1687:

> Because we perceive extension in all [bodies] that are sensible, therefore we ascribe it universally to all others also. That abundance of bodies are hard, we learn by experience; and because the hardness of the whole arises from the hardness of the parts, we therefore justly infer the hardness of the undivided particles not only of the bodies we feel but of all others . . . hence we conclude the least particles of all bodies to be all extended, hard, impenetrable, movable, and endowed with their proper inertia. And this is the foundation of all philosophy.

Lavoisier's philosophy of science was rigidly positivist, accepting only the numerical relationships revealed by experiment, rejecting atomism and mechanism alike as metaphysical, but his chemistry by contrast, brilliant as it was, could be dogmatic, and even infatuated on occasion—an affair of acid-formers and heat fluids. Newton's physics in the *Principia* was rigidly logical, admitting only the minimum of a priori constraints, yet his philosophy of science, even as expressed in the *Principia,* had atomism and the mechanical hypothesis for its foundation. No branch of modern science would appear to be more closely linked to the corpuscular and mechanical hypotheses than electrochemistry and the theory of solutions, yet even in its beginnings, Faraday struggled to clear it from its hypothetical entanglements.

The equations of Debye and Hückel, perhaps the most sophisticated expression of the theory to date, are in fact inspired by corpuscular reasoning; but the purist, aware of the dangers of too literal an extension from the world of the senses to the world of the ultimate particles, may possibly be excused for continuing to regard the equations as complete in themselves; equivalents, proportions, primes, expressing all of the facts of solution behavior, and rejecting all hypotheses.

10

The Periodic Law

THE PERIODS OF THE ELEMENTS HAVE THUS A CHARACTER VERY DIFFERENT FROM THOSE WHICH ARE SO SIMPLY REPRESENTED BY GEOMETERS. THEY CORRESPOND TO POINTS, TO SUDDEN CHANGES OF THE MASSES, AND NOT TO A CONTINUOUS EVOLUTION.

—MENDELÉEV, 1889

One view of the history of chemistry would be that of a continuing struggle between the chemical and the physical imaginations—between the analytic and the synthesizing mind. Once Kepler had looked for Platonic harmonies in the planetary orbits and Newton had asked whether the eye was contrived without skill in optics, or the ear without knowledge of sounds. "The distances traversed during equal intervals of time," Galileo wrote, "by a body falling from rest, stand to one another, in the same ratio as the odd numbers beginning with unity." The world was a complex of number and form, God's glory to conceal and man's to reveal—worship in the truest sense.

But as science hardened in the middle of the nineteenth century, becoming increasingly the province of professionals—successful men whose intelligence created wealth and power—its tolerance for the older, teleological methods declined. By the second half of the century the twenty-four elements in Lavoisier's textbook had increased to more than fifty. Far from the one "Catholick or Universal Matter" of Boyle, the complexity of the elements was becoming as great as the diversity of materials which chemistry was originally devised to explain.

When an earth once thought to be a simple element undergoes electrolysis, it is not resolved into the original four or five elements or even Lavoisier's twenty-four, but a new element may appear—yttrium or titanium—and on inquiry the chemist tells us that this is as far as analysis may go. We see small gain in this. It is as if the physicist, confronted with velocity, acceleration, momentum, and perhaps another fifty such, rather than resolving these into mathematical combinations of mass, length, and time, were instead to give us for each a separate term and announce that thus far and no further can analysis go. In the sixteenth century organic matter was composed of earth, air, fire, and water in various proportions. In the mid-nineteenth century it was composed of carbon, hydrogen, and oxygen in various proportions. Practically, there is indeed a great difference. Intellectually, one may at least sympathize with critics who saw little advance.

Attempts to organize the relatively chaotic assemblage of elements, to find meaningful connections and groupings among them, go back to Lavoisier's text. Misled by his estimate of the universality of the oxidation principle, Lavoisier grouped his elements according to how well they followed his chemistry rather than according to how well his chemistry described nature. After Richter's Rule and the proposals of Dalton and Avogadro, and the regularities discovered by Gay-Lussac, William Prout (1785–1850), a London physician, proposed in 1815–16 that hydrogen was the one fundamental element and that all the atomic weights were integral multiples of the weight of hydrogen. Although this hypothesis attracted support for a time, the temper of chemistry was strongly opposed. Lavoisier with his seven decimal places had set the stage for early-nineteenth-century chemistry, which was to remain, under the direction of Berzelius, Davy, Gay-Lussac, Liebig, and Wöhler, mistrustful of hypotheses.

With improved experimental techniques, results came closer and closer to the exact integral combining volumes of Gay-Lussac. But with the same improved experimental techniques, results diverged more and more from the integral proportions that Prout's hypothesis demanded. Other regularities appeared, notably the valency numbers of Frankland, which made chemical similarities—for example, between fluorine and chlorine, or between barium and strontium—intelligible. W. Odling (1829–1921), Secretary to the London Chemical Society, tried to arrange the elements in groups which would form a natural classification. Of his first or fluorine group (fluorine, chlorine, bromine, iodine) he wrote in 1857:

> These four substances have one marked property in common, a property not pertaining to any other element with which we are acquainted, namely, that of combining with hydrogen in the proportion of atom to atom, gaseous atomic volume to gaseous atomic volume. The combinations moreover take place without any condensation . . .

Odling classified the elements in thirteen groups of three, *triads* with similar properties:

> In each triad, the intermediate term is possessed of intermediate properties, and has an exactly intermediate atomic weight . . .

In one example:			
	Lithium	.. 6.5	
	Sodium	.. 23.0	$\frac{68.5}{3} = 22.8$
	Potassium	.. 39.0	
	Sum	68.5	Mean Difference = 16.

Intuitively the chemists sensed numerical regularities in the properties of the elements. J.A.R. Newlands (1838–1898) criticized the proposal of a correspondent that the atomic weights of the elements must turn out to be multiples of eight, substituting instead his own observation that the *differences* in the atomic weights of "certain allied elements . . . were generally multiples of eight . . ." or, taking the weight of oxygen as 16, ". . . the equivalent of oxygen is the unit of these differences. . . ."

The earliest periodic classification, however, was not primarily derived in this way. Beguyer de Chancourtois (1820–1886) was the successor to Élie de Beaumont (1798–1874) in the geology course of the Paris École des Mines. De Beaumont's ideas on the associations of elements in rocks, ore deposits, and the fluid emanations of volcanoes led de Chancourtois (1862) to the conviction that ". . . the distribution of the elements in the crust of the globe must in the last resort, be the basis of the classification of these elements . . ." The classification at which he arrived was based on a helical graph constructed on a cylinder, the *vis tellurique,* or *telluric screw.* Each turn of the screw was divided by sixteen equally spaced points, "characteristic numbers," corresponding to unit equivalent weight if, following Prout, hydrogen was taken as the unit. Differences of sixteen in equivalent weight brought elements into vertical columns; for example, sulphur with a weight of 32 appeared above oxygen on the second thread of the helix. Selenium and tellurium, with mineralogical properties analogous to sulphur and weights nearly 80 and 128, appeared on the fourth and seventh threads directly above oxygen; bismuth, 208, on the thirteenth. In the same way, Li, Na, K, and Mn fell in a single column or line cutting the helix on the side of the cylinder directly opposite the oxygen-bismuth column. Other helices of different pitches could be drawn on the cylinder between elements playing a mineralogically similar role, such as sodium and calcium in the feldspars. De Chancourtois believed that all the helices, "an infinity of helices of different pitches," through any two characteristic number points, "passing through others or even passing near, make the relationships of properties of a certain kind evident; analogies or contrasts show themselves through certain numerical orders of succession such as direct sequence, or alternating with various periods."

Struck by the numerical relationships everywhere in evidence, and particularly by "the predominance of the number 7," de Chancourtois spoke of his series as "essentially chromatic." The same Pythagorean analogy was the basis of Newlands' Law of Octaves.

Newlands was so confident of his numerical relationships that he did not hesitate to predict the equivalent weights for elements not yet discovered. By 1864 Newlands was proposing ". . . a law according to which the elements analogous in their properties exhibit peculiar relationships, similar to those subsisting in music between a note and its octave." The elements were arranged in the order of their equivalents, and, almost in passing, Newlands assigned to each a serial number ". . . calling hydrogen 1, lithium 2, glucinium 3, boron 4, and so on (a separate number being assigned to each element having a distinct equivalent of its own . . .)." By 1866, adopting the atomic weights of Cannizzaro, he had prepared the following table for a Thursday evening meeting of the London Chemical Society:

Elements arranged in Octaves

	No.		No.		No.		No.		No.		No.		No.		No.
H	1	F	8	Cl	15	Co & Ni	22	Br	29	Pd	36	I	42	Pt & Ir	50
Li	2	Na	9	K	16	Cu	23	Rb	30	Ag	37	Cs	44	Os	51
G	3	Mg	10	Ca	17	Zn	24	Sr	31	Cd	38	Ba & V	45	Hg	52
Bo	4	Al	11	Cr	19	Y	25	Ce & La	33	U	40	Ta	46	Tl	53
C	5	Si	12	Ti	18	In	26	Zr	32	Sn	39	W	47	Pb	54
N	6	P	13	Mn	20	As	27	Di & Mo	34	Sb	41	Nb	48	Bi	55
O	7	S	14	Fe	21	Se	28	Ro & Ru	35	Te	43	Au	49	Th	56

"Professor G. F. Foster humorously inquired of Mr. Newlands whether he had ever examined the elements according to the order of their initial letters?" This perspicacious reception of the periodic law was preserved for all time in the report of the meeting in the *Chemical News.*

In 1869 Dimitri Mendeléev (1834–1907), professor at St. Petersburg, undertook to write a textbook of chemistry and, to quote his account, "had to make a decision in favor of some system of elements in order not to be guided in their classification by accidental, or instinctive reasons."* Out of the requirements of teaching, a synthesis was forced on the recalcitrant world of chemistry.

Mendeléev's first table arranged some sixty-two elements in the order of their atomic weights, beginning with Li and starting a new period with every eighth element. Under this scheme each eighth element fell into a group with similar chemical properties. In the third period, calcium, Mendeléev's sixteenth element (not counting hydrogen), clearly fell in his

* Essentially the same work was independently performed by Lothar Meyer (1830–1895), the German chemist who was also responsible for the adoption of Avogadro's hypothesis.

second group, with beryllium and magnesium. But his next heaviest element, titanium, had properties that placed it in the fourth group, with carbon and silicon. Like Odling—but unlike Newlands, who had placed chromium immediately following calcium—Mendeléev recognized a gap. "Vacant places occur for elements which perhaps shall be discovered in the course of time," he wrote; moreover, "in this manner it is possible to foretell the properties of an element still unknown." Making use of an approximate rule of Döbereiner (1780–1849), Mendeléev averaged the weights of the four elements adjacent to the gap—two neighbors in the period and two in the group—to obtain the predicted atomic weight of the missing element:

> For example, selenium occurs in the same group as sulphur, S = 32, and tellurium, Te = 125, and, in the 7th series As = 75 stands before it and Br = 80 after it. Hence the atomic weight of selenium should be ¼ (32 + 125 + 75 + 80) = 78, which is near to the truth. Other properties of selenium may also be determined in this manner. For example arsenic forms H_3As, bromine gives HBr, and it is evident that selenium, which stands between them, should form H_2Se, with properties intermediate between those of H_3As and HBr. . . . *In this manner it is possible to foretell the properties of still unknown elements.*

Mendeléev predicted the properties for *ekasilicon,* Es (discovered and named germanium, Ge, by Winckler in 1887), which would fall between silicon and tin in the fourth group and between zinc and arsenic in the fifth period.

> Its atomic weight is nearly 72 [today 72.60], higher oxide EsO_2, lower oxide EsO [wrong], compounds of the general form EsX_4, and chemically unstable lower compounds of the form EsX_2. Es gives volatile organo-metallic compounds—for instance, $Es(CH_3)_4$, $Es(CH_3)_3Cl$, and $Es(C_2H_5)_4$, which boil at about 160°, etc.; also a volatile and liquid chloride, $EsCl_4$, boiling at about 90° and of specific gravity about 1.9 . . . the specific gravity of Es will be about 5.5, EsO_2 will have a density of about 4.7, etc.

The specific gravity of germanium is 5.5, of GeO_2 4.7; the compound $Ge(C_2H_5)_4$ boils at 160° C.; the liquid chloride $GeCl_4$ has a boiling point of 83° C. and specific gravity of 1.9.

With the same success, Mendeléev described the unknown elements now known as gallium and scandium:

> . . . when I conceived the periodic law, I deduced such logical consequences from it as could serve to show whether it were true or not. . . . No law of nature can be established without such a method of testing it. Neither De Chancourtois, to whom the French ascribe the discovery of the periodic law, nor Newlands, who is put forward by the English, nor L. Meyer, who is now cited by many as its founder, ventured to foretell the *properties* of undiscovered elements. . . .

In this manner, with the periodic law, the first phase of the development of classical chemistry was complete. A set of empirical laws—the law of equivalent weights (Richter), of multiple proportions (Dalton), of integral volumes (Gay-Lussac), of molecular constitution (Avogadro), of valence (Frankland), and of serial number (Newlands)—were summarized in Mendeléev's periodic law. Each element was now characterized by a set of numbers, its equivalent weight, its valence(s) or group number, its period, and finally its serial (atomic) number. Although these numbers were not obtained without the intervention of atomic and molecular hypotheses, they were logically independent of such hypotheses. All, including Avogadro's number, could reasonably be termed constants, appearing in experiment and conveying no information other than conjectural about the ultimate constitution of things.

We have seen that Dalton's reading of Newton was in error yet led him to assert the law of multiple proportions on what was at first a single case and one which he misinterpreted (in part) at that. Gay-Lussac refused to accept Avogadro's hypothesis, and if by the sixth decade of the nineteenth century it was used at all, it was as a convenience in calculating the atomic weights—a number not representing the number of molecules in unit volume of a gas, as Avogadro had hypothesized, but rather a number which, if assumed, would yield a single set of equivalent weights, rather than the many sets which chemistry brought to the Karlsruhe Conference.

At this time, then, atomism was a generally accepted concept, widely used as a convenience in the visualization of difficult problems in chemistry (stereochemistry, for example), and already so woven into the fabric of chemistry that only by a conscious effort could the terms of atomism—terms, said Faraday, "which express the hypothesis as well as the fact"—be avoided. Nevertheless, Faraday was not alone in rejecting the hypothesis. The attitude went back to Lavoisier, himself an atomist, who yet rejected atomism in the conviction that it was metaphysical; a crutch, said Pierre Duhem, for the weak but ample English mind—and to be avoided by the precise Gallic guardians of true science.

11

The Idea of Energy and the Assault on Materialism

. . . WE MAY CONCEIVE THE SYSTEM CONNECTED BY MEANS OF SUITABLE MECHANISM WITH A NUMBER OF MOVEABLE PIECES, EACH CAPABLE OF MOTION ALONG A STRAIGHT LINE, AND OF NO OTHER KIND OF MOTION. THE IMAGINARY MECHANISM WHICH CONNECTS EACH OF THESE PIECES WITH THE SYSTEM MUST BE CONCEIVED TO BE FREE FROM FRICTION, DESTITUTE OF INERTIA, AND INCAPABLE OF BEING STRAINED BY THE ACTION OF THE APPLIED FORCES. THE USE OF THIS MECHANISM IS MERELY TO ASSIST THE IMAGINATION . . .

—MAXWELL, 1873

Two fundamental changes in the nature of science occurred during the nineteenth century. The first was the dematerialization of the element fire, which by the late eighteenth century had become heat and the element called caloric, and in the twentieth century is merged into the energy concept. It was a major blow to materialism and a major part of the second fundamental change, which was the emergence of a new kind of science, abstract in the extreme, mathematical in its reasoning, positive in its assertions and its conclusions. Thermodynamics, the science of heat and work and energy, was one of the results.

To turn the heat fluid, the tiny corpuscles of fire, into something so abstract as a flow of energy and as intangible as work, in fact an equivalent to work, was to attack materialism at its base and to substitute for the Democritean unity of atoms in the void, a new duality of matter and energy. As there was a rigid law of the conservation of matter, the nineteenth century established the law of the conservation of energy and elevated it to a canon of classical science. In place of the material kinds of matter, such as CO_2 gas, which could enter into compounds and be obtained again from them, keeping their mass while completely transforming other properties, the "forms" of energy, such as heat or the motion of a material body, were not in themselves conserved. Rather, work was transmuted into heat, or electricity into motion, with definite ratios of exchange, which however did not hold in reverse.

Finding the laws governing these transformations turned out to be the equivalent of inventing new, immaterial but mathematically precise *things*—concepts called *entropy* or *chemical potential*. It was a science as far removed from Kekulé's benzene ring of whirling snakes as his structural formulae were from the alchemist's pentagon. But if it was a revolution, it was like all revolutions—with its roots in the immediate past, specifically in the seventeenth and eighteenth centuries' triumphal harnessing of the power of fire to do useful work.

The steam engine appeared during the eighteenth century, and with it a glimpse of an almost limitless Baconian development of society. Although some of the benefits of science and the new chemistry were widely diffused by 1900, it is doubtful that the alchemy of plastics, cermets, semiconductors, fiberglass, and solid fuels was foreseen even then. But after the steam engine, one could glimpse a future in which the muscle power of man and animal would be completely replaced by machine and the obligation laid upon Adam of earning his bread by the sweat of his brow would be removed. Steam power was foreseen by Hooke in 1679 as another outcome of the long chain of research that led from Galileo's consideration of the problem of the pump through Torricelli and Viviani, to von Guericke, to Boyle's Oxford chambers. Hooke proposed obtaining a vacuum by condensing steam in a closed vessel. Such vacuums were tried by Thomas Savery in 1698 for raising water in mine shafts. Denis Papin, a Huguenot refugee who shared Hooke's quarters at Gresham College when he first escaped to England, suggested that vacuum could be transmitted along a pipe. Papin made a pressure cooker and designed a steam gun, while Hooke nearly drowned in a diving bell of his own invention.

By 1712 Thomas Newcomen (1663–1729), an English blacksmith who probably corresponded with Hooke, had made the first practical steam engine. Its operation depended largely on the change in volume when water changes from the liquid to the vapor state, a change greater than twelve-hundred-fold. The steam in a cylinder was condensed by cold-water jets and the pressure of the atmosphere pushed a piston down on the Hooke vacuum in the cylinder. More steam was admitted to the cylinder, raising the piston, and the cycle repeated. The alternate lowering and raising of the piston operated the pump rods in a mine. In order to keep from turning his fire on and off, Newcomen made the steam in a separate boiler and used valves to open and close the pipe from the boiler to the cylinder.

Newcomen's, the first practical steam engine, was successfully operated at many of the English coal mines—the same coal mines that made the industrial revolution in England. Yet on close analysis it could easily be seen to waste fuel. The vast, poorly fitted brass hulk of the cylinder and piston were alternately heated and cooled in every cycle, but useful work was performed only through the expansion and contraction of the steam.

James Watt, who began as an instrument maker in the laboratories of the University of Glasgow, observed that the small-scale models of Newcomen engines consumed disproportionately large quantities of fuel for the trifling amounts of work produced. Watt consulted with Joseph Black, who was teaching at Glasgow and was concerned with the concepts of heat capacity and latent heat.

Black, who later followed Lavoisier in abandoning phlogiston and replacing it with caloric (conceived as a material substance and in fact the second element in Lavoisier's *Treatise*), discovered that substances differed in their capacity for heat:

> ... very soon after I began to think on this subject, (*anno 1760*) I perceived ... that the quantities of heat which different kinds of matter must receive, to reduce them to an equilibrium with one another, or to raise their temperature by an equal number of degrees, are not in proportion to the quantity of matter in each ... This opinion was first suggested to me by an experiment described by Dr. Boerhaave (Boerhaave, Elementa Chemiae, exp. 20, cor. 11). After relating the experiment which Fahrenheit made at his desire by mixing hot and cold water, he also tells us that Fahrenheit agitated together quicksilver and water unequally heated. From the Doctor's account, it was quite plain, that quicksilver, though it has more than 13 times the density of water, produced less effect in heating or cooling water to which it was applied, than an equal measure of water would have produced.*

The same amount of fuel which will heat a single gram of water from 0 to 100° C. will heat about ten grams of iron through the same temperature range. Defining the calory as the quantity of heat required to raise the temperature of one gram of water from 14.5 to 15.5° C., the heat capacity for iron in the same range is 0.116 cal. $gm^{-1}deg.^{-1}$. In these experiments Black also discovered that very large quantities of heat (still thinking of heat as substance) were hidden or stored in the process of changing from solid to liquid and from liquid to vapor. In melting one gram of ice, 79.7 calories are absorbed—enough heat to raise the temperature of the same gram of water from 0 to nearly 80° C. To vaporize one gram of water at 100° C. requires 539.6 calories. These large quantities of heat do not raise the temperature of the gram of new steam. It is still at 100°—the same temperature as the water at the start. To raise the temperature of the steam and cause it to expand requires additional heat above and beyond the 539.6 calories latent in the steam.

Black's was not the only way in which heat could have been defined. C.H.A. Holtzmann (1811–1865) pointed out in 1845 that besides raising the temperature of matter, an inflow of heat would "increase the elasticity" of a gas or do mechanical work "the equivalent of the elevation of temperature."

* Black, 1803, p. 184.

> The heat can only be measured by its effects; of the two effects mentioned the mechanical work is the best adapted for this purpose, and it will accordingly be so used in what follows. I call the unit of heat the heat which by its entrance into a gas can do the mechanical work *a*—that is, to use definite units, which can lift *a* kilograms through 1 meter.

Watt's analysis convinced him that the useful work of the engine was performed in the expansion and contraction of the envelope of steam. The repeated heating and cooling of the cylinder in the Newcomen engine and the changing of state from liquid to vapor and back to liquid were losses of heat reducing the power and efficiency of the machine. He was able to eliminate the first loss by employing a separate condenser kept at a temperature lower than 100° C. The working cylinder and piston were then kept at a temperature higher than 100° C., and needed to be heated only once when the engine was started. Steam condensing in the separate condenser produced a partial vacuum which was led by pipe *à la* Papin to the cylinder. With lower pressure in the cylinder the pressure of the atmosphere forced the piston down (closing the valve to the condenser at the bottom). Then live steam from the boiler was admitted and the piston driven up again.

The picture of Watt the mechanic, "hard-nosed" man of affairs, with no more time for science or thought than a childish familiarity with his mother's tea kettle, while dear to the hearts of the modern Goths, is hardly borne out by this history. We have already met Watt as an associate of Joseph Priestley and Erasmus Darwin in the Birmingham Lunar Society; dissenters in religion, republicans or at least liberals in a monarchy, men of learning in a mob-ridden society. Science, it has been said, owes more to the steam engine than the steam engine owes to science. James Watt would hardly have accepted this as a compliment, or indeed as anything other than the patronizing dismissal which it is.

Others besides Watt, notably John Smeaton (1724–1792), applied themselves to the improvement of the steam engine. In 1824 a young French engineering officer, Sadi Carnot (1796–1832), set the problem in the most general terms and began a

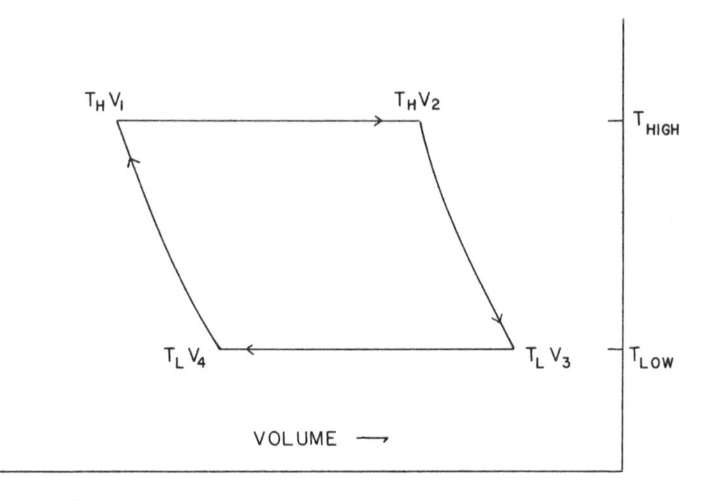

Figure 13. The Carnot Cycle—Temperature vs. Volume

theory of heat and work (thermodynamics) which by the end of the nineteenth century stood as a model for all science. Carnot abstracted the problem from the consideration of any particular kind of heat engine and restricted his attention to the flow of heat and work performed under ideal circumstances. He conceived of an ideal engine (the Carnot engine) operating in a cycle (the Carnot cycle). As the gas in a real engine expanded, pushing a piston and therefore performing work, its temperature fell from the temperature of the boiler T_H to that of the condenser T_L. This expansion Carnot divided into two ideal steps. In the first, the gas is in contact with the high temperature source of heat. This step is *isothermal;* that is, because it is in contact with the source, the temperature remains unchanged. As the gas expands from V_1 to V_2, doing work on the machinery connected to the piston, heat flows into the system to keep the temperature constant. In the second step the temperature falls from the temperature of the source to that of the condenser, or *sink*. The temperature cannot fall to a lower temperature than that of the condenser or sink because heat does not of itself flow except from a higher to a lower temperature. In Carnot's analogy the flow of heat is compared to the flow of water in a water wheel. The water cannot fall below the general level of the stream bed without having to be pumped out again. Although the system does more work on the outside machinery, no heat flows because the system is not connected to the source. The gas expands from V_2 to V_3. Such an expansion without heat flow is called *adiabatic.*

In the third step of the Carnot cycle the steam, now in thermal contact with the low-temperature sink, is compressed isothermally to V_4. Work of compression is performed by the outside machinery on the gas, and since compressing a gas would heat it were it not in contact with a suitable sink, this heat flows out of the gas into the sink. In the fourth and final step, the gas is disconnected from the sink and compressed from V_4 back to the original volume V_1, adiabatically. The volume V_4 was chosen so that adiabatic compression to V_1 would raise the temperature back to the temperature of the source T_H. Work of compression is again required to accomplish the last step. But this work of compression is just sufficient to raise the temperature of a fixed amount of gas by $T_H - T_L$ degrees and is therefore equal to the work done by the expanding gas in the second step, when its temperature fell through $T_H - T_L$ degrees. The work required in the last step is the work supplied in the second. In the first step the gas expanded isothermally at a high temperature and therefore a high pressure. The amount of work performed on the outside in the first step was greater than that required to compress the gas in the third step at the lower temperature of the sink. Adding the heat and work balance for the four steps, we get (counting + as into, or on the gas, and — as out of, or by the gas) $+ Q_1 - W_1 - W_2 - Q_3 + W_3 + W_4$ with $W_1 = W_4$. At the end of the cycle the system is back at V_1T_H, having performed $W_1 - W_3$ units of work on the outside, while Q_1 units of heat flowed into the system from the source and Q_3 units were rejected by the system to the sink.

Here Carnot's reasoning became quite close:

> . . . We may justly compare the motive power of heat with that of a fall of water. The motive force of a waterfall depends on

the height and the quantity of fluid: the motive force of heat depends upon the quantity of caloric employed and what we may call the height of its fall, that is to say, the difference in temperature of the bodies between which the caloric is exchanged.

This is the theorem of Carnot. If the temperature of the sink is almost as great as the temperature of the source, so that the pressure required in compression in step three is almost as great as the pressure obtained in expansion of step one, W_1 will be little different from W_3 and it will require as much work to return the piston to its starting position as the piston can perform. Clearly the greater the "fall of caloric" the greater the amount of work gained in the cycle. At the same time the amount of work performed is also dependent on the quantity of the gas undergoing the transformation, so that Carnot's principle is logically firm.

But nothing has been said about the nature of the gas, whether it be steam or air or mercury vapor; and nothing has been said about the mechanics of this engine, whether internal combustion or steam or gunpowder, whether reciprocating or jet or turbine. A moment's reflection will show that for the words "cylinder" and "piston" used in the description of the cycle, we could substitute some such phrase as "appropriate mechanical connection" with no change in the validity of the argument. Carnot's engine is an ideal engine, a "thought" engine, and therefore as superior to a real engine as thought always is to things. It is not a perfect engine, because its efficiency is determined by the difference in temperature between the source and the sink. Since the sink can be no colder than the outside surroundings, some heat must always be rejected at a finite temperature. Such heat is no longer available for useful work. This is like saying that ultimately no waterfall can operate below sea level.

To prove this theorem Carnot suggested that his engine could be operated in reverse. The same engine operating between the same limits would go through the same steps but in the opposite order. At the end of a cycle the situation would be as it was at the beginning of a cycle. The universe would be unchanged except that now heat would have been absorbed from the sink and rejected to the high temperature source, that is, heat would have flowed from the colder to the hotter body with the expenditure, rather than the gain, of work $W_1 - W_3$. Misled by the concept of a material caloric, Carnot thought that, as in a water engine, the water could not be increased or destroyed, so the same amount of caloric must be rejected as was absorbed. Therefore, he reasoned, no heat engine could be more or less efficient than another.

He proved this by assuming the opposite. If a better engine could be made, it would be able to obtain more work from the flow of the same amount of caloric through the same temperature differential. Let the more efficient engine drive the less efficient engine in reverse, while Q units of heat tumble T degrees down the temperature scale. The less efficient engine at the same time raises Q units of heat back up the temperature falls to the temperature of the source, where they are ready to drive the "good" engine through another cycle. But the "good" engine does more work than the "bad" engine operating with the same Q units of heat through the same temperature difference. This excess work is available for outside use. At the end of the double cycle the universe is unchanged except for the appearance of this extra work. This is perpetual motion of the second kind—so called because it is obtained through a violation of the second law of thermodynamics.

> . . . there would thus result not only the perpetual motion but an indefinite creation of motive power without consumption of caloric or of any other agent whatsoever. Such a creation is entirely contrary to the ideas now accepted, to the laws of mechanics and of sound physics; it is inadmissible.

It is inadmissible, Carnot says. It would be obtaining something for nothing. It would be miraculous, or at least magical, to obtain an effect out of all proportion to its cause. We object to this for the same reason that we object to gambling. Nature must be consistent. God is not mocked. The universe must be an orderly system operating according to inexorable law or else there is no purpose to existence. The Herr Gott does not play at dice, Einstein said. Without law, without purpose, our lives are meaningless and empty. We become nothing but animals, the creatures of blind chance at the mercy of capricious and unpredictable elements—devils—in a world without God.

The reader will have perceived that Carnot has not performed a physical experiment to prove the underlying assumptions of Greco-Judaic thought, but rather, the heir to two millennia of Western civilization, he has proved his theorem by deduction from the concept of order. There is a fitness to things as they are, a sense of proportion. Carnot has assumed the conservation of energy.

His theorem is based on the impossibility of perpetual motion of the second kind, that is, the production of useful work simply by the rearrangement of energy. Something is gained in his cycle—useful work. Something is lost—heat has flowed from the source to the sink. Under ideal conditions the amount of work gained would be just sufficient to operate the Carnot engine in reverse and return the original amounts of heat from the sink back to the source. But any departure from perfection in the operation of the engines will mean that not quite enough work will be gained and less heat can be brought back. The best that could be accomplished would be to restore the original state of the universe. All leaks in the machinery, all friction, all mistakes will conspire to prevent this ideal restoration.

Carnot also violated the same law of conservation of energy which he had assumed. The quantity of heat rejected to the sink is not the same as the quantity accepted from the source. It is less by the amount of work done. The false trail of the Paris chemists, Lavoisier, Laplace, Fourcroy, had at last come to an end. Caloric was not a material substance.

Benjamin Thompson, the New Hampshire schoolmaster who became Count Rumford and hired Humphry Davy for the Royal Institution, early redirected the attention of science to the old Baconian concept of heat as a fine motion of particles. Thompson noticed the prodigious quantities of heat produced in boring brass castings for cannon. He measured the heat capacities of cast brass and of the borings which had already yielded heat enough to boil many times their weight of water. "The capacity ought not only to be changed," he wrote, "but the change undergone . . . should be sufficiently great to account for all the Heat produced." There was no change in the heat capacities of the brass before and after boring.

> It is hardly necessary to add, that anything which any *insulated* body, or system of bodies, can continue to furnish *without limitation,* cannot possibly be a *material substance;* and it appears to me to be extremely difficult, if not quite impossible, to form any distinct idea of anything, capable of being excited and communicated in these experiments, except it be MOTION.

The same idea impressed itself upon the mind of J. R. Mayer (1814–1878), a ship's physician of Heilbronn, who had noticed that in the tropics the venous blood of the ship's passengers remained red, while that of the hard-working sailors turned blue. Knowing that the red color of the blood was due to oxygen, Mayer reasoned that in the tropics the bodily processes of combustion consuming the oxygen in the blood were retarded, except in the case of the sailors who used the energy for work. Heat—caloric—was one of Lavoisier's elements of chemistry.

> . . . we find that in all chemical operations [wrote Mayer]—combinations as well as decompositions—changes of temperature occur, which . . . are of all degrees of intensity from the most violent heat downwards. We have measured quantitatively the heat developed, or counted the number of heat-units, and have so come into the possession of the law of the evolution of heat in chemical processes.
>
> We have long known, however, that in innumerable cases heat makes its appearance where no chemical action is going on; for instance, whenever there is friction, when unelastic bodies strike one another, and when aeriform bodies are compressed.

From published values of the heat absorbed by expanding gases, Mayer computed the amount of heat equivalent to a given amount of work. Heat and work were forms of something more general and fundamental—*energy*—and the forms of energy were interconvertible one into the other, just as solids and liquids and gases were forms of something more general—matter. Mayer proposed as a fundamental law of physics that energy can neither be created nor destroyed. The heat which appeared in Rumford's brass borings had its origin in the work of the boring. Unable to dissipate the chemical energy of oxidation in their bloodstreams, his ship's passengers in the tropics accumulated oxygen in their venous blood.

Suspicious of metaphysics, Poggendorff's *Annals of Physics* rejected Mayer's paper, but Liebig published it in his *Journal of Chemistry* in 1842. James Joule (1818–1889), son of a Scottish brewer, published works on the electrical equivalent of heat as early as 1840. Joule took Rumford's measurements as the starting point for very careful experimental determinations of the exact amount of work equivalent to a unit of heat. The Royal Society showed no more patience with Joule's experiments than German physics had shown with Mayer's theories. In 1847 Joule, reporting his measurement of the mechanical equivalent of heat, was asked to be brief. This was the meeting at which the twenty-three-year-old William Thomson (1824–1907), later Lord Kelvin, first grasped the significance of the conservation law—the principle of Joule which was to become for Kelvin and for nineteenth-century science the single great fundamental truth of nature.

Kelvin also discovered the work of Carnot, which had been applied in engineering physics by his classmate Clapeyron. There appeared to be a contradiction between the principles of Carnot and Joule. According to Carnot, it was the flow of heat from source to sink that produced the work. But according to Joule, heat would have to be *converted* into work. There could be no question that Carnot was right and that heat had to be rejected to the sink, but on the other hand it was equally clear that Joule was right and that energy had to be conserved. For some time Kelvin pondered the apparent paradox. He used the Carnot cycle as the basis for an absolute temperature scale, one which would not depend upon the properties of any particular substance, such as the thermal expansion of mercury, for its validity. Equal units of temperature on the Kelvin scale were separated by differences between which an ideal Carnot engine would yield equal amounts of work. Carnot himself had made notes on the conservation and interconvertibility of energy shortly before his death from cholera in 1832. "According to certain ideas that I have formed on the theory of heat, the production of a unit quantity of motive power necessitates the destruction of 2.7 units of heat."

The great difficulties involved in accurate quantitative measurements of heat had prevented him from realizing in 1824 that the quantity of heat rejected to the sink was less than the quantity accepted from the source by the amount of the work produced. This was clearly pointed out in a critical memoir on Carnot's paper by Rudolf Clausius (1822–1888):

> A careful examination shows that the new method does not stand in contradiction to the essential principle of Carnot, but only to the subsidiary statement *that no heat is lost,* since in the production of work it may very well be the case that at the same time a certain quantity of heat is consumed and another quantity transferred from a hotter to a colder body, and both quantities of heat stand in a definite relation to the work that is done.

Clausius showed that in rejecting the idea of a substantial caloric one must also reject the concept of the total heat of a body—that is, the concept that if we specify the temperature, pressure, volume, mass of a body, a single number will represent the units of heat in a body. Referring to the diagram on page 186, there are many different pathways by which we might bring a body into a given condition—some involving work, some a flow of heat, and some a combination of the two. We may speak of the algebraic sum of the heat and work expended in the process by which the body reached its state—the *energy* of the body—but it is no more proper to refer to the heat in the body than to the work in the body. Like work, heat is a process, it is energy in transit. This idea Clausius termed the first law of thermodynamics. The energy of the universe is a constant. It is expressed as $U = Q + W$, with U standing for total energy and Q and W for heat and work respectively. Further, Carnot's principle tells us that there is a function (Carnot's function) constant for all substances at the same temperature, representing the maximum work to be derived from a given heat flow. Kelvin put it:

. . . the second fundamental proposition is completely expressed by the equation

$$\frac{dp}{dt} = \mu M,$$

where μ denotes what is called "Carnot's function" [ratio of the work to heat per degree] . . . Also, we see that Carnot's expression for the mechanical effect derivable from a given quantity of heat by means of a perfect engine in which the range of temperature is infinitely small, expresses truly the greatest effect which can possibly be obtained in the circumstances; although it is in reality only an infinitely small fraction of the whole mechanical effect of the heat supplied; the remainder being irrecoverably lost to man, and therefore "wasted," although not *annihilated*.

The previous year Clausius had noted Kelvin's objection to the Carnot theory, that while heat must flow through a temperature differential to produce work, it was in no wise true that every such flow of heat produced work: "Heat can be transferred by simple conduction, and in all such cases, if the mere transfer of heat were the true equivalent of work, there would be a loss of working power in Nature, which is hardly conceivable."

Although the loss or destruction of working power in nature was inconceivable, its wastage came to occupy Clausius' attention. In place of Kelvin's μ, representing the limit to the efficiency of a thermal engine (process), he derived a similar but simpler function measuring the degradation of working power in nature, or the unavailability of energy for work. This function he called *entropy* and defined mathematically.*

Carnot's principle is still correct. Although by the principle of Joule, energy is conserved, Carnot's principle tells us the conditions under which heat may be converted to work. At the end of a Carnot cycle, the working body (steam) is in the same state as at the beginning, therefore its internal energy is unchanged. The entropy function, Clausius showed, depended only on the actual condition of the body considered, and not,

* The total of all changes in the internal energy or "latent heat" of a body and the external work performed on or by the body, divided by the temperature at which the changes take place.

as in the case of heat and work, on the means by which the body reached its condition. If two ideal Carnot engines were operated, one operated in reverse by the other, at the end of a cycle—the working body (steam) being returned to its original condition and the heat which flowed from source to sink in the first engine being restored from sink to source in the second—the total change in entropy of the system would be O. The work performed by the first engine was just sufficient to drive the second engine in reverse, completely restoring the original state of the universe. If in place of an ideal engine we substitute a totally "bad" one, in which the heat flow occurs but no work is produced because of friction, leaks, and so on, then the entropy increases sharply: X calories flowed through the engine from the source temperature T_H to the lower temperature T_L without accomplishing the work of driving the second engine in reverse and restoring the original state of the universe. The source is less able to do work. Energy has not been lost; it has flowed to the sink. The unavailability of the energy of the universe has been increased. Since all real engines, which is to say all heat processes, must of necessity be less than perfect, entropy in real processes must always increase. Clausius expressed it as, "The entropy of the universe is always increasing."

With the concept of entropy we reach a new level of scientific sophistication. Heat and temperature are concepts which we intuitively grasp, but the idea of a physical quantity invented for its mathematical convenience was a relatively new one in 1850, the year of Clausius' first memoir. Heat is something real —manifest to the senses. Entropy, we say, is something unreal— an exercise in the differential calculus. Yet it is not necessarily true that we sense directly things like heat or temperature. Our Aristotelian scientist working on common sense would fail to conceive of the quantities of heat that may be pumped out of well water to warm homes in the Tennessee Valley. The tap water which feels cold to someone inside the house on a winter day feels hot if he has been handling snow without gloves. Do we sense directly mass—or its resistance to change of motion which we call inertia? We certainly do not sense the atom or the electron. Entropy has actually more reality than these, since these are hypothetical models of reality, while entropy is an invented quantity and should rather be compared to a number or a geometric form.

A more fruitful way of examining our concepts might be *operational.* According to the operationalist, the concept is synonymous with the set of operations used in its measurement. The concept of temperature is on the simplest level, synonymous with the readings which measure the expansion of a column of mercury. The concept of the electron is the swing of a galvanometer plus the division of certain numbers by other numbers. For the operationalist, concepts or ideas which are not measurable in themselves are meaningless. He is like the preacher who was asked what God was doing before he created the heavens and the earth, and replied that he was busy preparing hell for people who asked such questions.

It might be better to speak of observables and to consider their degree of direct measurability. Length would be a first-order observable; temperature, since it is measured by a change in volume or length or color, would be a second-order observable. On this scheme both heat and entropy would be higher-order observables, both escaping the outer darkness of the operationalist, since the operations for the determination of both quantities may be rigorously defined.

In yet another sense the thermodynamics which began with Carnot illustrates the development of a new sophistication in chemistry. For although the new mathematics had entered physics in the seventeenth century, it was to enter chemistry only in the nineteenth. To read Newton's *Principia,* plane geometry albeit of a high complexity, was sufficient. But already in the works of Lagrange in the eighteenth century, and especially of Laplace, who with Lavoisier had conceived of caloric, Newton's mechanics were changed so as to be unrecognizable in the glitter of their new mathematical cloaking. The parallel development of physics and mathematics had far outdistanced chemistry.

As early as 1657, a quarter-century before the *Principia,* Fermat (1601–1665) had anticipated both the integral calculus and the conception of new physical quantities out of mathematics, with his "principle of least time." In the eighteenth century Maupertuis (1698–1759) defined *action;* a concept extended by Euler and Lagrange to a first law of dynamical nature. Before entropy, mathematical analysis through differential equations and the postulation of general laws in the form of functions (Gauss' *constraint,* Hamilton's function, the Laplacian, and so on) had become the standard method of the physicist.

One required little more than arithmetic, however, to read the works of Dalton and Gay-Lussac, while Lavoisier and his predecessors were largely qualitative. In the first stage of chemistry a burning splint is observed to flare brightly in a stream of gas which then is labeled oxygen. In the second stage, weights and volumes are the significant concepts and numbers are manipulated. In Joseph Black's account of specific heat, he finds it unnecessary to employ even algebraic equations. Carnot was a product of the École Polytechnique, a school which his mathematician father had been influential in founding. He was trained in the method of mathematical reasoning which had brought France to the foremost position in science. But in his memoir of 1824 Carnot's analysis was largely verbal, with only an occasional resort to algebraic equation. The expression of propositions in the form of differential equations appears only in the footnotes. A quarter-century later Clausius presents a verbal analysis and goes forward to assign differential expressions to the verbal concepts with which he deals. From these the operations of the differential calculus yield new relationships which on translation back to the phenomenological world (the world of things and machines) predict the outcome of thermal processes.

In Kelvin's μ and Clausius' later invention of entropy we see for the first time in chemistry the emergence of concepts out of the requirements of mathematics which enter into science and rapidly demonstrate their power in elucidating the workings of nature. Some examples of these concepts of thermodynamics are entropy and enthalpy, the Gibbs and Helmholtz functions (free energy), the chemical potential. All of them express a law or theorem in their conceptualization—as the defining differential equation for entropy is a statement of the second law:

$$dS = \int \frac{dQ}{T} \geqq O; \text{ S is entropy; Q is heat; T is absolute temperature.}$$

Like the systems of dynamics, such as Laplace's "Celestial Mechanics," which had appeared before in physics, thermodynamics is a set of mathematical principles—laws governing the relationships of heat, work, and energy. Just as Newton's mechanics were the necessary consequences of the three fundamental laws of inertia, momentum, and action-reaction, thermodynamics is the consequence of the two fundamental laws. If we consider a trivial chemical reaction

$$H_2 + \tfrac{1}{2}O_2 = H_2O + Q$$

the first law does not tell us which way the reaction will proceed or indeed whether or not it will proceed. The chemical A need not necessarily combine with the chemical B. Berthelot's assumption that the reaction would "go" in the direction that would evolve heat is completely wrong, since many reactions occur which absorb heat. Nor is the problem trivial, as the occasional explosion of a munitions or fertilizer works proves.

The first law enables us to balance the energy economy of the reaction. The energy of the explosion of the plant could be calculated nicely. It is the second law which tells us that H_2 and $\tfrac{1}{2}$ O_2 go to form water and that the oceans will not suddenly separate into masses of hydrogen and oxygen with the absorption of all of the earth's heat. It is the second law which enables us to design a chemical plant so that reactions proceed in the appropriate directions. The second law tells us the direction of things—the direction of entropy increase. It introduces time into science, which would otherwise be outside of history. It introduces a necessary asymmetry into nature. For if all the world were to run backwards, the earth's rotation and revolution to reverse, rivers run uphill, and old men turn into children, Newtonian mechanics would not be confounded and only the second law would remind us that our film was running backwards and that time has a unique direction.

As Newton was obliged in the *Principia* to define his terms carefully—terms of difficulty such as absolute time and space, or mass and momentum—so thermodynamics is obliged to define its terms—equilibrium, temperature, heat. In the rigorous form of the late eighteenth century, Newton's laws appear as equations of the differential calculus—relationships between the fundamental physical quantities. In the simplest sense, if enough of these quantities are known, then the equations are solved—numbers are found for each of the unknowns. But the solutions for the differential equations of motion are themselves differential equations. In the end, every mechanical problem can be reduced to the solution of such an equation or set of equations. In thermodynamic problems we require the *state* of the system—values for the functions such as temperature, pressure, energy, entropy, known as the thermodynamic coordinates because they locate the condition or *state* of the system in the same way that the geographic coordinates of latitude and longitude locate the position of a point in space. The interdependence of these functions are given by *equations of state,* such as the ideal gas law of Boyle and Gay-Lussac, $PV = RT$, or the amended version by van der Waals. Such equations of state are empirical, arising from experiment, and can in no way be derived from first principles, unless hypotheses about the structure of matter—the kinetic-molecular hypotheses, for example—are introduced.

In a typical thermodynamic process such as the expansion of a gas, we say that the state of the gas which is specified by its temperature, pressure, and volume undergoes a change. If we can measure any two of these, the value of the third is known from the equation of state. The two laws of thermodynamics define functions of state, energy, and entropy with values which depend only on the state of the system and not at all upon the processes by which it arrived at that state. The rules of the differential calculus applied to the defining equations tell us how the changes in one variable are related to the changes in others. It is these *changes*—the expansion of a gas against pressure, the flow of energy—which constitute work and heat and which the operations of the calculus relate in this way to the laws of thermodynamics.

In the case of the expansion of a gas in contact with the source of heat of a Carnot engine, we can calculate the actual flow of heat from the source. We obtain this from the laws which connect changes in entropy with changes in the coordinates P, V, and T. An equation of state enables us to eliminate one of the variables, pressure. The definition of entropy enables us to substitute another—heat—for entropy. We wind up with an equation specifying the exact flow of heat for a given expansion of gas in the first step of a Carnot engine.

From the two fundamental laws of thermodynamics, the operations of the calculus enable us to completely specify the values of all quantities of interest in the system. Just as Newton, from his definitions and from his three axiomatic laws of motion, could undertake, once given the position and velocity of a comet, to calculate directly its position and velocity at any time past or future; so the science of thermodynamics, given two of the three coordinates P, V, and T, and the appropriate equation of state, can completely determine the whole of the history, the whole of the future, of a thermodynamic system.

Thermodynamics does not in itself attempt to explain the mechanism of heat processes, although it is commonly used together with molecular theory. J. W. Gibbs wrote in 1901, ". . . We avoid the greatest difficulties when, giving up the attempt to frame hypotheses concerning the constitution of material bodies . . ."

Consider a solid transforming to a vapor. We obtain an expression connecting the change of pressure with temperature to the heat and volume changes. We obtain this by mathematical manipulation of the differential expressions defining a function of state, restricting our mathematics by the limitations of the first two laws. This tells us the necessary logical interrelationships between certain suitably defined entities, Q, T, P, V. The equation of state relates these to the observed interrelationships between physical quantities, essentially assigning physical significance to our logical abstractions and reducing the number of variables from three to two. If now in an experiment we measure one of the two—as for example the change of pressure with temperature through the transformation—arithmetic at once gives us the latent heat. If anything about this discussion suggests to the reader the parturition of an elephant resulting in a mouse, he might reflect upon the advantages in certain circumstances—as in the fueling of an aircraft or the designing of a nuclear power plant—of prediction of the magnitude of the reactions.

In effect, and for the chemist, thermodynamics is the science of determining in advance how large a boat may be built in the cellar. Like Newtonian mechanics it is a complete science, in that starting from a few axiomatic statements, the mathematical equations governing all thermodynamic processes may be derived. Newton's *Principia* contains *all* the necessary consequences of his three axioms. It contains, for example, the theorem already mentioned that "*if* the density of a fluid which is made up of mutually repulsive particles is proportional to the pressure, the forces between the particles are reciprocally proportional to the distance between their centers." The theorem is mathematically proved, therefore true. But it does not apply to gases, since as we know, gases do not fit the condition. They are not made of mutually repulsive particles. We must combine empirical relationships with Newton's mathematical principles of natural philosophy.

Newton's mathematics tells us that Kepler's laws are the consequences of an inverse square force of gravitational attraction—yet only an astronomer at a telescope can tell us that planets do exist and do move according to those laws which Kepler first revealed, and that therefore the inverse square law is indeed the true law of universal gravitation. So much, like Descartes, we can obtain a priori. But in the end, if our science is to be of the world and not of the mind we resort to the world to see that it is indeed the inverse square—and not, as it might equally well have been, the inverse cube. There is something outside of our mind, outside of man, outside of thought. In thermodynamics, the equations of state play the role of Kepler's laws and indicate that of all the infinity of possible ways in which a process might proceed with constant energy and increasing entropy, it does in fact proceed under given conditions according to one of these equations of state, rather than the others.

Historically, we see that the development of this logical science, whose ingenious proofs and reasoning draw from the admiring mathematicians the adjectives "elegant" and "beautiful," follows from the intensive development of mathematical physics. Carnot and Clapeyron were direct students of the great French tradition of mathematical physics, inspiring in turn, as this tradition spread to England, Kelvin and Maxwell; to Central Europe, Helmholtz and Clausius, later Boltzmann and Planck; and to America, Willard Gibbs—a list which should in justice be many times as long.

The kind of science, complete, precise, analytic, which had so excited the admiration of the Age of Reason had been extended to chemical process. The perfectly simple first and second laws were the axioms from which the whole complex science of heat and work could be derived. As in mechanics, no hypotheses were feigned; no statements about the ultimate constitution of matter, whether atomic or continuous, assumed. Thermodynamics knows nothing of the ultimate nature of heat or energy —whether waves, particles, the random motion of molecules, or mysterious fluids—as mechanics knows nothing of the ultimate nature of mass, or gravitation. "It is enough," Newton said, "that gravity does really exist, and act according to the laws which we have explained, and abundantly serves to account for all the motions of the celestial bodies, and of our sea." It is both the strength and the weakness of thermodynamics that it too is free of hypotheses. No future discoveries, no latter-day revelations of the nature of heat or matter, can shake the thermodynamic edifice.

Galileo's discovery that Jupiter had moons was sufficient to make untenable the idea that the earth was the center of the universe. Lavoisier's experiment with the calx of tin in a sealed vessel was sufficient to upset the doctrine of phlogiston. But for thermodynamics to be challenged it would be necessary to overthrow the first and second laws—for energy to be created or destroyed, for time to be reversed, or cold vessels spontaneously to raise their own temperatures. Never, it seemed, had the temple of science rested on so firm a foundation.

12

The New Mechanism

GIVEN FOR ONE INSTANT AN INTELLIGENCE WHICH COULD COMPREHEND ALL THE FORCES BY WHICH NATURE IS ANIMATED AND THE RESPECTIVE SITUATIONS OF THE BEINGS WHO COMPOSE IT—AN INTELLIGENCE SUFFICIENTLY VAST TO SUBMIT THESE DATA TO ANALYSIS—IT WOULD EMBRACE IN THE SAME FORMULA THE MOVEMENTS OF THE GREATEST BODIES OF THE UNIVERSE AND THOSE OF THE LIGHTEST ATOM; FOR IT NOTHING WOULD BE UNCERTAIN AND THE FUTURE, AS THE PAST, WOULD BE PRESENT TO ITS EYES.

—LAPLACE, 1819

After his doctoral studies at Zurich, Albert Einstein (1879–1955) was employed in the Swiss patent office at Berne. In 1905, in the famous volume XVII of the *Annalen der Physik,* Einstein published three papers. The last proposed a major revision of classical mechanics—the special theory of relativity. The first extended a new idea of Max Planck's about discrete bits of energy called quanta, to the theory of light. The second, with which we are here concerned, seems at first glance to be of a very different nature. It is as perfectly mechanical as the others are antimechanical. In place of the leaps of the imagination demanded by the theory of relativity, or the challenge to the wave theory of light in Einstein's extension of the quantum hypothesis, we are asked to be completely mechanical and to interpret all the properties of matter, in the manner of Hooke, as the results of the operations of tiny machines.

In 1905 this hypothetical fine structure of matter—the construction of models representing matter as made of particles which by their figure, number, and motions would explain all the properties of matter—seemed to be the ultimate goal of science. The kinetic-molecular theory, which explained heat as molecular motion, was the link by which classical mechanics was extended to cover all science. Joule's relationship between heat and work was simply a matter of scale. A body moves across a plane. Work is consumed; heat of friction appears. The gross motion of the body has been analyzed into a myriad of microscopic motions of the molecules of the body and the plane. The first law of thermodynamics in this view is simply an old law of Newtonian mechanics, the law of conservation of momentum.

At this time two very different views of classical science were prevalent. Pierre Duhem, uncompromisingly rigorous, espoused a scientific formalism which he declared, in the manner of Lavoisier, to be the truest expression of the French mind. He had been trained in thermodynamics, the "doctrine which is the paradigm of abstract theories." There was but one method of science for him, the method of axiomatics, of laying down exact postulates from which the operations of mathematics draw precise and necessary conclusions. Other methods he described as too simple, "too crude for a physics concerned with some degree of exactitude."

"There is no necessary connection between the inductive method canonized by Newton and the mechanistic conception of physics," Duhem wrote.

> If we look for what is common in the very numerous theories, and they are very disparate besides, brought together . . . under the name mechanism, this is what we find: All these theories seek to represent physical laws by means of groups of solid bodies with dimensions close to those that we can see and touch, that can be sculptured in wood or metal; whether they are formed of molecules or atoms, of ions or electrons, the systems whose motions the theorist describes are, despite their extremely small size, conceived of as analogous to majestic astronomical systems. All these speculations are alike, therefore, in the following: They wish to

reduce all the properties we observe in nature to combinations of shapes and motions subject to expropriation by the imagination.

Again and again in the history of science the fundamentally Aristotelian viewpoint appears. Reason, logic, the foundations of the beliefs on which our society and the social order rest, compel us to this formalistic view of knowledge. It is nature alone which, deaf to reason, refuses to be constrained to our sensible requirements. The light and heavy stones refuse to fall as Aristotle prescribes. Through Galileo's telescope we see that Jupiter and Saturn are themselves circled by satellites, that there are other centers to the motions of stars than our earth. Atomism and mechanism may offend us by the impurity of their intellectual origins, but it is nature's, not our prerogative to prescribe. If Duhem could cite with approbation Newton's famous disclaimer, *I frame no hypotheses,* Newton would hardly have applied the same stricture to God. Astronomers, including Copernicus and Galileo, sought to constrain the motions of the planets to perfect circles, but it was Kepler's triumph that he reversed this problem and, rather than fit the observations of planetary motions to our human concepts of geometric form, chose to find the geometry which fit the observed motions of the planets. It is not our metaphysics which determines nature, but rather nature which sets the limits within which our understanding strives. Few individuals have better understood this than Albert Einstein.

Einstein proposed in his second 1905 paper that Avogadro's number, and with it all related atomic and molecular quantities, could be determined by an analysis of the motions of visible particles suspended in a liquid. (This was simultaneously proposed by M. V. Smoluchowski [1872–1917], then a student under Loschmidt's successor.) Such motions of dust particles in air and pollen dust in liquid had been familiar since they were first observed by Robert Brown in 1828. Einstein reasoned that "a dissolved molecule is differentiated from a suspended body *solely* by its dimensions." But dissolved molecules have *macroscopic* properties, that is, properties which are directly observable and measurable. They exert pressure (osmotic) and they diffuse through the liquid according to well-known laws. Duhem was writing at the same time:

> Abstract thermodynamics furnished Gibbs from the start with the fundamental equations for [osmotic pressure]. Thermodynamics has also been the sole guide of J. H. van't Hoff in the course of his first works, while experimental induction furnished Raoult with the laws necessary for the progress of the new doctrine. The latter had reached maturity and constitutional vigor when the mechanical models and kinetic hypotheses came to bring it the assistance it did not ask for, with which it had nothing to do, and to which it owed nothing.

From thermodynamics it was already known that the macroscopic properties of matter such as pressure, temperature, specific heat, and volume do not depend upon the sizes of molecules or their equivalent weights but upon their molarity (concentration of equivalent weights). "Why should not . . . a number of suspended particles . . . produce the same osmotic pressure as the same number of molecules?" The visible motions of the suspended particles are subject to the laws of mechanics. Einstein treated complex motions of a very large number of visible particles by statistical analysis to arrive at their "pressure" and "temperature" and other aggregate properties. In effect, Einstein showed the connection between these aggregate or averaged-out mechanical properties of visible and therefore directly observable bodies with the macroscopic properties of solutions, and thereby implied that solutions had a molecular fine structure. If the only difference to be found is one of size, then what indeed is to prevent us from representing the laws of thermodynamics "by means of groups of solid bodies . . . that can be sculptured in wood or metal"?

"If the movement discussed here," wrote Einstein, "can actually be observed . . . then classical thermodynamics* can no longer be looked upon as applicable with precision to bodies even of dimensions distinguishable in a microsope: an exact determination of actual atomic dimensions is then possible."

The observations of Jean Perrin (1870–1942) and others on

* Specifically the law of increase of entropy of a system (see pages 194–195).

Brownian motion soon confirmed the theory. Values for the Avogadro number gradually converged on 6.06×10^{23} (now 6.02×10^{23}). This determination by Einstein and Perrin has been called the Magna Charta of molecular physics. After this Wilhelm Ostwald (1853–1932) and Ernst Mach (1838–1916), the last important holdouts against atomism, were convinced. Exact values could now be placed upon the weights and sizes of atoms, the charge of an electron, and the interatomic distances in matter.

It is remarkable that in his first paper, before admitting that Brownian motion was indeed the thermal motion with which he was concerned, Einstein had the figure 6×10^{23} for the Avogadro number, while Perrin's determinations for a number of years were low by about 10^{23} particles. It was not the determination of the number that convinced Ostwald and Mach, but rather the discovery of the same mathematical relationships for effects such as diffusion and osmotic pressure for aggregates of particles large enough to be visible, as were known for molecular solutions. Avogadro's number, in the guise of "Loschmidt's number," had been first determined after 1865, when Loschmidt applied the results of the newly emerging statistical mechanics to determine the size of a molecule of gas.

The relationships between the diameter of a molecule, the average distance it would travel before colliding with another molecule, and the number of molecules, had already been established by Maxwell and Clausius, although of the three numbers, only the second, the average free path between collisions, could be calculated without experimental data. Assuming that the molecules of a gas are pressed together when the gas is condensed to a liquid, Loschmidt found a connection between the volumes of gases condensed to liquids and the numbers of their molecules. To actually count the number of molecules he required the condensation coefficients for ordinary gases such as oxygen and nitrogen, which had not at the time been liquefied.* These he found following an ingenious method

* The mean free path—known—is a function of the size and number of particles (Maxwell and Clausius); the condensation coefficient—observable—

used by Hermann Kopp (1817–1892), an early-nineteenth-century historian of chemistry. Kopp had been involved in the attempt to extend Gay-Lussac's law of combining volumes to solids and liquids. From tabulations of specific volumes of carbon-hydrogen compounds, C_6H_{14}, C_7H_{16}, C_8H_{18}, etc., it was possible to find numbers for the carbon atomic volume and the hydrogen atomic volume which, multiplied by the 6, 7, and 8, etc., of the formulae, would fit the observed specific volumes. Using these atomic volumes and the same procedure with carbon, hydrogen, oxygen, and nitrogen compounds, it was possible to find numbers which would give a reasonably consistent agreement with observation. Loschmidt's method contained too many uncertainties and approximations for accuracy, but his figures yield values about one order of magnitude from the modern value. Most of the error was in his method of calculating condensation coefficients.

It was after this work of Loschmidt's, and in the period of its greatest triumphs, that the atomic theory fell into disrepute. The excesses of Descartes, who imagined screw-like corpuscles to account for magnetism, should have been compensated by the ingenious devices of his follower Huygens, who accounted for the double refraction of light in Iceland spar by postulating elliptically shaped corpuscles and with this simple model accounted for the optics as well as the crystallography of the mineral. Huygens' optics, his constructions for the behavior of light in crystals, the geometry of his wave theory of light, are in use today.

Daniel Bernouilli, one of a distinguished family of mathematicians driven to refuge in Switzerland by the massacres of the Protestants in the lowlands, had derived Boyle's law of the elasticity of gases from the simplest assumptions of the mechanical theory: that an "elastic fluid" consists of "very minute corpuscles, which are driven hither and thither with a very rapid motion." Considering that the average velocity of the corpuscles would be unchanged when the gas was compressed

---

is a function of the size and number of particles (Loschmidt); therefore the system of two equations with the two nonobservables—distance and number—is completely determined.

to half its former volume, Bernouilli reasoned that the number of collisions of the gas with its container would be doubled, the "repeated impacts" constituting pressure. But reasoning of this sort offended the sensibilities of the new French mathematical physicists such as Lagrange, who boasted that his treatise of analytic geometry contained not one single diagram—respect for abstract mathematics had gone so far. Dalton and Gay-Lussac had both rejected Avogadro's hypothesis, although for different reasons.

Joule, in 1857, was the first to calculate an absolute molecular quantity. He preferred the atomistic hypothesis of Bernouilli, which had been revised by John Herapath (1790–1868), to the Cartesian concepts of continuity revived by Humphry Davy. Assuming the mass of a gas contained in three heavy particles, moving parallel to the three edges of a cube, Joule set the problem of finding at what speed such particles must move in order to produce by repeated impacts on the wall of the container the observed pressure of one atmosphere. Noting that the force of a single impact is proportional to twice the mass of the particle times its velocity, and that the number of impacts is also proportional to the velocity, Joule wrote, "It is manifest that the pressure will be proportional to the square of the velocity of the particles." N particles of the same total mass as three heavy particles moving at the velocity v would produce the same observed pressure, and therefore our velocity, considered as an average, is independent of the determination of N. For hydrogen, at 0° C., Joule found a velocity of 6,055 feet per second which compares with a modern value of about 5,620 feet per second. Knowing the dependence of the pressure of a gas on temperature, Joule was able to compute the absolute 0 of the temperature scale at 491° F. below the freezing point of water* (or −273° C.) and to relate his mechanical equivalent of heat to the specific heat of the gas.

James Clerk Maxwell (1831–1879) was Faraday's intellectual heir. He supplied the sophisticated mathematics which Faraday lacked, seizing upon Faraday's results, casting them in terms of purest mathematical abstraction, and deriving from them

* A figure previously obtained by Herapath, according to Maxwell.

immense scientific profit. It was Maxwell who turned Faraday's field model into the system of equations which make up the electromagnetic field theory and the classical theory of light.

Intrigued by Clausius' attempt to find the average distance traveled by a molecule between collisions, Maxwell saw that this would be related to properties of a gas, such as its internal friction and its heat conductivity. If the flow of heat were to consist of an exchange of velocities of colliding particles, the conductivity of heat would be related to the number of collisions. "These phenomena," Maxwell wrote, "seem to indicate the possibility . . . In order to lay the foundations of such investigations of strictly mechanical principles, I shall demonstrate the laws of motion of an infinite number of small, hard, and perfectly elastic spheres acting on one another only during impact." The laws of motion for such impact were known, but Maxwell proposed to determine the laws for the molecular model of a gas, that is, for a very large number of very small bodies in random motion. Although he could do this only statistically, if the statistical laws were referred to a large enough sample—a large enough number of particles—they would be highly accurate. Further, the statistical laws of motion that Maxwell proposed to obtain would predict the macroscopic behavior of gases. "If experiments on gases are inconsistent with the hypothesis of these propositions, then our theory, though consistent with itself, is proved to be incapable . . ."

Beginning with hypotheses concerning the ultimate nature of matter, Maxwell by strictly mathematical deduction drew from these hypotheses equations representing the observable behavior of matter. Whether matter really behaved as predicted could be settled only by resort to experiment. In his first paper on the subject, in 1860, Maxwell actually made mathematical slips which he and Boltzmann later corrected. Yet the mathematical soundness of such a derivation is not the proof of the molecular model. We remember that Newton was mathematically sound in his model of gases, but his model was totally wrong. Ultimately it is nature which makes the decision, although only temporarily. For nature indeed at first made the decision for phlogiston, as nature decided for Dalton in favor of Newton's erroneous model. We may say with confidence that experiment may break a theory, but no theory is ever confirmed for long. Whatever the state of our science may be, we can be sure of one thing and one thing only, that it is not permanent truth and that it can and will be changed.

Maxwell and Boltzmann obtained an equation giving the proportion of particles whose velocities lie between given limits (the law of distribution of velocities among the molecules). The Maxwell-Boltzmann law is therefore the *velocity distribution law.* It predicts, as one might expect and as later experiments have verified, that the largest number of particles travel at velocities near the mean; that the farther away a given velocity is from the mean velocity, the fewer the particles with such a velocity. Since the lowest velocity possible is 0 and there is no limit to the highest possible velocity, the curve of the frequency of distribution of the velocities is skewed or lopsided in the direction of 0.

Suppose that a large stream of particles all traveling at the same speed are suddenly admitted to a box. These particles colliding with one another and with the sides of the box would rapidly exchange speeds, until after a sufficient number of collisions a small number would have acquired greatly increased speeds, a small number would have lost most of their initial speed, and the largest number would be traveling at speeds close to the initial, or average, speed. The Maxwell-Boltzmann distribution represents the most probable distribution, or the equilibrium distribution. When numbers of particles of the order of 10^{23} are involved, any departure from the regularities of the Maxwell-Boltzmann curve must be rapidly erased. If two cubes of gas, one at $0°$ C. and one at $100°$ C., are placed side by side, and the partition between them is removed, the velocity-distribution curve representing both cubes at the instant of removal would have two humps for the average speeds of the molecules at the two temperatures. But a moment later the faster molecules from the hot box would collide with the slower molecules of the cold box, and the resultant exchange of velocities would bring the velocity-distribution curve

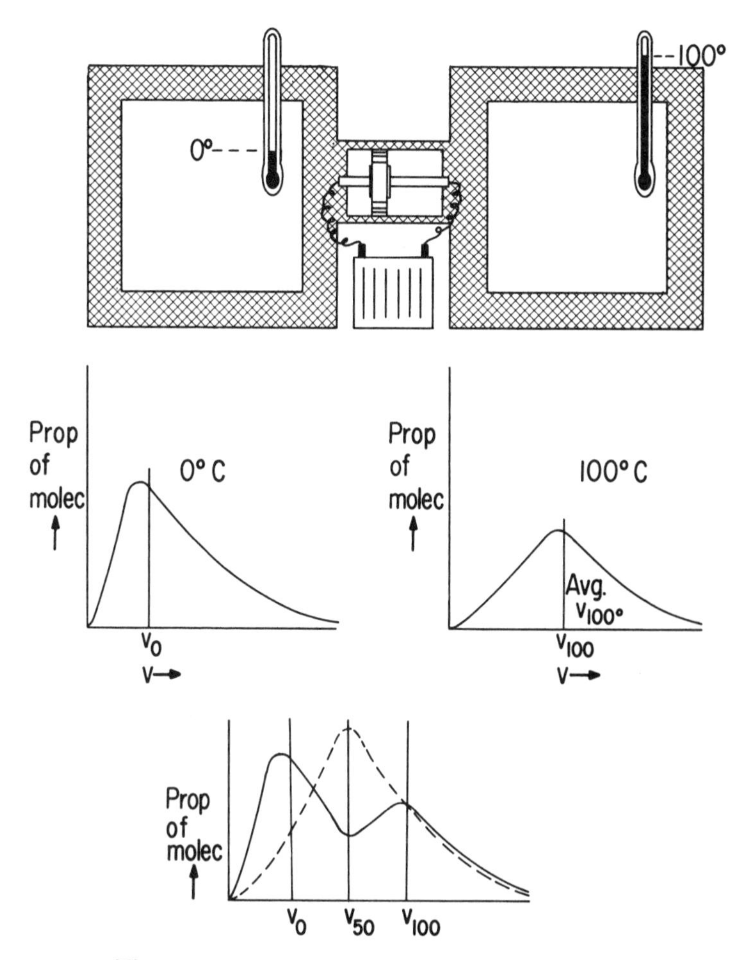

Figure 14. The Experiment of the Two Cubes
The Carnot cycle explained in kinetic-molecular terms.

back to the curve of the Maxwell-Boltzmann distribution, with its single hump at the average speed characteristic of 50° C. The double-humped curve is highly improbable. It corresponds to a sampling of seawater in which we find at one spot fresh water and, nearby, salt water. If such a situation exists momentarily, when a bucket of fresh water is emptied into the sea, it is immediately erased by ordinary thermal agitation. The uniformly salt sea is the most probable state. We do not expect the salt of the Atlantic to concentrate momentarily in the Arctic Sea, leaving the remainder of the sea fresh, although physically this is one possible way of distribution of the salt in the sea. There are, however, so very many more ways in which the salt may be distributed to produce a uniform salinity that any departure from this salinity we unhesitatingly ascribe to an extraneous cause—a river mouth, an iceberg, a current.

> We can calculate the state of equilibrium [Boltzmann wrote] by calculating the probability of the different possible states of the system. The initial state will in most cases be a very improbable one and from it the system will progress toward more probable states, until it at last reaches the most probable state . . . in accordance with the second law this change must always occur in such a way that the total entropy of all the bodies increases; according to our present interpretation this means nothing else than that . . . the system of bodies goes from a more improbable to a more probable state . . .

Boltzmann derived his transport equation in 1872. It is a differential equation, with solutions which specify the state of a system of particles under given conditions. From this equation he derived a function which he called the H-function, summarizing the probabilities of the particle distribution. Boltzmann's H-theorem asserts that this function must decrease* in any spontaneous process such as the mixing of systems of particles. It plays the same role for idealized particle systems as the negative of the entropy for thermodynamic systems. This H-theorem extends the kinetic-molecular hypothesis to all of thermodynamics. The artificial assumption of the second law, the mysterious entropy and the arbitrary assertion of its increase, is replaced by the logic of probabilities. The formalism

* The value of H decreases with time until, after the system of particles reaches the Maxwell velocity distribution, it remains constant. For example, the largest proportion of the particles will be traveling with velocities as shown by the peak of the curve (Figure 17b), while very few will be traveling at exceptionally higher velocities. All the particles will continually change their velocities, but the *proportions* traveling at two velocities, and therefore the probabilities of the distribution, are unchanged.

of the *energeticists*—Ostwald, Mach, Duhem, Helmholtz—was undercut.

Even "the fundamental equations for the motion of individual molecules will turn out to be only approximate formulas which give average values, resulting . . . from the interactions of many independent moving entities . . ." Boltzmann wrote. As Gibbs put it:

> The laws of thermodynamics . . . express the approximate and probable behavior of systems of a great number of particles, or, more precisely, they express the laws of mechanics for such systems as they appear to beings who have not the fineness of perception to enable them to appreciate quantities of the order of magnitude of those which relate to single particles.

Josiah Willard Gibbs (1839–1903) was America's first theoretical physicist. It was Gibbs in his statistical mechanics who most perfectly reflected the Euclidean *apagoge,* the method of axiomatics. Not only could all the principles of thermodynamics be derived from the first principles of mechanics, but in Gibbs' work, and in the work of Boltzmann, there was a general method which would prove applicable even when the necessity arose for changing the first principles of mechanics (the quantum theory). Unlike Pierre Duhem, neither Gibbs nor Boltzmann allowed metaphysics to direct their scientific thought. Both were concerned with physics first, even in the creation of their masterpieces of pure analysis. Gibbs wrote:

> The first problem of molecular science is to derive from the observed properties of bodies as accurate a notion as possible of their molecular constitution. The knowledge we may gain of their molecular constitution may then be utilized in the search for formulas to represent their observable properties.

J. D. van der Waals (1837–1923), who had obtained a greatly improved version of the ideal gas law by allowing for intermolecular forces, was Gibbs' example.

Gibbs' method was to represent the state of a physicochemical system by analyzing it to a great number of particles differing only in their positions and velocities. In the mechanistic atmosphere of the latter nineteenth century any system could be so represented, as Maxwell indicated (page 181), by the use of a suitable and purely imaginary mechanism. For example, the expansion of a gas could be represented by increasing the average velocity of ideal mass points (molecules). The specified configuration and velocities of the whole system are its *phase*. It is a Laplacian microcosm, its past and future completely determined by the $6n$ numbers, three for the position and three for the momentum of each of its n particles. Just as two numbers fix the position of a point on an ordinary two-dimensional graph, and three numbers specify a point in ordinary three-dimensional space, the $6n$ numbers specifying the phase of the system locate a point in what Gibbs called the *phase space*.

"Let us imagine a great number of independent systems, identical in nature, but differing in phase, that is, in their condition with respect to configuration and velocity," he wrote, designing an *ensemble* of systems.

In this $6n$-dimensional phase space, the points, each representing one of the systems, form a kind of gas or molecular fluid. The laws governing the evolution of this gas as the points stream through the hyper-space are the subject of Gibbs' investigation. Under certain restrictions the points behave like an incompressible fluid. The number of points occupying a given element of hyper-volume in the space is its density, and the probability that a given system is in a given state is governed by this *density in phase*. Gibbs defined an index of this probability by its natural logarithm. He was able to show that this probability depended upon the energy associated with the configuration and velocity of the systems of the ensemble. The distribution of the systems in phase space which conformed to their probabilities, he called the *canonical ensemble*.

In this fashion Gibbs devised a statistical mechanics. Its assumptions are the laws of Newtonian mechanics. It says nothing about real systems except as they may be represented by Gibbs' abstractions. Even the coordinates of thermodynamics are abstracted; in place of entropy (S) we have the index of probability (ln P). In place of heat (Q) we have the flow of energy (dU). In place of temperature we have the proportion between the index of probability and the energy. Nor is the method restricted to those ideal systems, all identical in nature, which Gibbs called *petits ensembles,* or even to the ensemble of different equilibrium systems—the grand canonical ensemble—all of which might be assumed to be covered by the elementary laws of mechanics, by Newton's laws, for example, statistically applied. The imaginative conceptions of Gibbs proved the starting point for the understanding of intra-atomic phenomena and of the departures from classical behavior on the part of the fundamental particles themselves. "And this would be true if the ensemble of systems had a simultaneous objective existence. But it hardly applies to the creations of the imagination."

Statistical mechanics began with the atomistic speculations of Hooke and Boyle and Newton. Their particles or corpuscles were simply smaller replicas of shapes that could be sculptured in wood or metal. While these naïvely simple particles were evolving into atoms, molecules, and even the electrons and nuclear particles of today, mathematicians were demonstrating that the averaging of the motions of particles governed by Newtonian mechanics led to the laws of physical chemistry. Bernouilli in the eighteenth century derived Boyle's Law ($PV = k$) from the motions of ideal mass points. A century later van der Waals, by allowing for the size and even chemical character of the molecules, obtained the first of the modern equations of state. These explanations of the properties of gross, sensually apparent matter as the averaged-out behavior of bits of the same matter differing only in their size, culminated in the papers of Einstein and Smoluchowski on the Brownian motion.

At the same time, as more detailed knowledge of the phenomena for which the hypothetical particles were conceived accumulated, it became clear that they did not always behave in exactly the fashion of gross particles. Molecular vibrations, for example, do not cease at the absolute 0 of temperature; the electronic oscillations in which light originates do not on the average absorb equally the available energy; the mass of the heavy atom such as gold is not uniformly distributed through the space which it occupies. Gibbs wrote:

> Difficulties of this kind have deterred the author from attempting to explain the mysteries of nature, and have forced him to be contented with the more modest aim of deducing some of the more obvious propositions relating to the statistical branch of mechanics. Here, there can be no mistake in regard to the agreement of the hypotheses with the facts of nature, for nothing is assumed in that respect.

We discover these things by experiment with large quantities (ensembles) of the hypothetical particles, and the laws of mechanics which we construct to describe the outcome of our experiments are statistical in turn. Rather than predict with Newton the return of a comet, a single particle constrained by the laws of motion to a future completely determined, we predict with Fermi or Born the distribution curve for the electrons among the energy states in the heart of a distant star. All our experiments involve great numbers of the hypothetical particles, and the observations that are the outcome of our experiments are statistical averages. When we realize this—realize, for example, that the high intensity of the component of the Roentgen radiation excited from a metal is due to the very large number of the electronic interchanges occurring—we see that our task is not simply the prediction of a result of an experiment, but the prediction of the statistical distribution of the results of a large number of individual experiments. A great number of electrons bombard an Avogadro's number of copper atoms, inducing all possible electronic interchanges within the atoms and the radiation of energy in all possible frequencies. If at a particular frequency the intensity of the radiation is great, we say that for a large number of atoms the statistical incidence of that particular interchange is great, or for the individual atom the probability of that particular interchange is high.

In an ideal experiment a ray of light falls on two closely spaced slits in a screen. The pattern which appears on a photographic plate beyond the screen is not one of two bright lines, but of a single bright line at the center of the pattern and a series of discrete lines of lower intensity on either side. If a single *photon* (particle) of light approaches the screen, through which of the two slits will it pass? In general, we cannot say. Nor can we say where on the plate it will land. What is worse, after it has passed through the screen and its position has been recorded, as for example by a counter tube, we cannot say through which of the two slits it passed. We can only say that the probability of its landing at the center of the pattern is greatest, and in fact the probabilities are directly related to the intensities of the previous experiment. Let a large number of photons—a ray of light—fall on the screen, and the new mechanics, a statistical mechanics resting on the new axioms of the quantum theory, will predict the resulting pattern of distribution.

If this procedure, abandoning the determinism of scientific law and limiting one's aims to a determination of the probabilities, recognizing even in these a fundamental uncertainty, would not have satisfied the mechanists of the seventeenth century, neither did it satisfy Einstein, who sought in the theory of the Brownian motion a demonstration of the reality of atoms and molecules.

Boltzmann too, although he called himself an idealist, considered that his methods confirmed the objective reality of the molecular theory. In his later life the kinetic theory was attacked as the crude approximations of weak minds. Boltzmann sank into a deep melancholy. Could it have been the new discoveries—radioactivity, relativity, the quantum hypothesis of Max Planck—contradicting the laws of thermodynamics, challenging even Newtonian mechanics, which defeated this man whose life had been devoted to the construction of a theoretical buttress for classical science? How could Boltzmann, who had himself predicted that even the axioms of Newton would eventually be shown to be statistical averages, have failed to foresee the extension of his science to the new world of phenomena revealed by the genius of Einstein and Planck? "I am conscious of being only an individual struggling weakly against the stream of time. But it still remains in my power to contribute in such a way that, when the theory of gases is again revived, not too much will have to be rediscovered," he wrote. Eight years later he killed himself.

13

The Revolt in Physics and the Chemical Bond

THE WORLD'S BASIC PRINCIPLES ARE VIOLATED, THE CONSTITUTION OF THINGS IS OVERTURNED, THE MYSTERIOUS OPERATIONS OF NATURE ARE ABORTED, THE HERDS OF ANIMALS ARE SCATTERED, ALL THE BIRDS CRY OUT IN THE NIGHTTIME . . .

—CHUANG TZU, 300 B.C.*

With the suicide of Boltzmann foreshadowing events to come, science crossed the divide from the tidy, cultivated garden of classical thought to a new thicket of stubborn, irreducible fact. Once again, as in the period which culminated in Aristotle and the Academy, the exercise of pure reason had brought about a complete system which was now challenged. The older scientists of the second half of the nineteenth century felt—quite prematurely, as it turned out—that they had succeeded in plugging the dangerous gap of experiment through which so much novelty had poured irretrievably since the seventeenth century. In a kind of fusion of idealism and positivism, the Austrian physicist-philosopher Ernst Mach proposed that science was to be considered as a logical system striving not so much toward objectivity as for simplicity and expedience (economy) of thought. He rejected the "reality" of phenomena and in particular of matter (substance) and atomism, considering with Lavoisier that these were meta-

* Quoted in Creel, 1963.

physical questions. For Mach the senses were the only source of physical concepts, and science was the organization of these into rigidly logical systems.

Ironically, it was Einstein himself who, in his 1905 paper on special relativity, most successfully employed Mach's method. Einstein proposed to replace the classical concepts of motion, resting as they did on the laws of Galileo and Newton, with two new principles, one empirical, the other epistemological. The first was the principle that measurements of the velocity of light in vacuum were always constant, regardless of the motion of the observer or the source. The second asserted that the laws of physics were to be cast in such a form as to be the same on bodies in uniform motion as on bodies at rest.

To reconcile the two principles required that the concepts of space and time, and with them of mass and energy, be reformulated. In place of fixed values and laws of conservation (the principles of Lavoisier and Joule) length, time, mass, and energy were relative measures, taking on values according to the relative motion of the observer and the system observed, values which would maintain the constancy of the velocity of light and satisfy the relativistic equations which were the new fundamental laws of physics. In its emphasis on observation as the root of law and on the invariance of scientific law, the special theory of relativity was a direct example of Mach's thought, although Mach himself was completely hostile to it.

In 1915 Einstein's General Theory of Relativity appeared, with even more violent challenges to classical thought. The basic properties of matter, mass, and inertia Einstein asserted were completely equivalent and experimentally indistinguishable. Space and time were linked to form a continuous *field* interacting with matter. The field alone, like the tracks in a freight yard, governed the motion of the objects within it. The universal, simplest laws of physics were the equations of the geometry of the field.

A still more radical challenge to classical concepts had been thrown down by Max Planck (1858-1947) in endeavoring to account for the color of the light emitted by an incandescent solid, that is, a solid heated to the point of giving off light. According to the electromagnetic theory of Maxwell, light was a wave with length of the order of 10^{-5} cm. (or 1/100,000th of a cm.) and frequency of the order of 10^{5} (or 100,000) cycles per second. Heinrich Hertz (1857–1894) had demonstrated that the rapid oscillation of an electric charge was the origin of such waves, and it was assumed that some such vibration of the electrons within the atom itself was the source when light was given off by a heated body. Since the light emitted by heated solids contained all possible frequencies (colors), the atomic oscillators were evidently capable of absorbing and emitting a reasonably continuous and complete spectrum.

The theory of the emission spectrum of an ideal solid was developed by J. W. Strutt, Lord Rayleigh (1842–1919), and James Jeans (1877–1946) at the turn of the century. This emitted light was called "black body" radiation because a perfectly black radiator will absorb and emit light of every color equally well. The theory led to a most curious paradox which Ehrenfest named "the ultraviolet catastrophe." Jeans imagines a hollow box with a hole in it. The inside walls of the box are atoms and therefore atomic oscillators. Energy is put into the cavity, for example, by shining a light into the hole, or for a real body, by heating it to incandescence. The energy excites the oscillators making up the cavity walls, each oscillator in turn, as it begins to vibrate, emitting light which excites other oscillators, so that the light is reflected back and forth within the cavity. Presumably the vast number of atomic oscillators will reach an equilibrium, receiving and emitting equal energies on the average. If now an opening is made in the cavity from which light streams forth, what should be the frequency of the light? There would appear to be no reason for one color to be favored over any other, and we should expect the cavity to give off white light, which includes all the colors. Heating the cavity should make the light more intense, but offhand we can see no reason for temperature to affect the color. Actually, the black body radiation is a dull red at the lowest temperatures (525° C.) at which visible light is emitted, shifting through a cherry red to an orange-yellow to a blue-white at about 1200° C. Our expectations of uniform whiteness at all temperatures are directly contradicted by experiment.

A string stretched between two posts is a reasonable model of an ideal oscillator, from which we can derive the possible ways of vibration (called *modes*). The string is free to vibrate everywhere except at the posts where it is tied. It therefore may vibrate with any wave whole multiples of which will fit between the fixed ends. If the length of the string is L, the possible waves will be of lengths 2L/1, 2L/2, 2L/3, 2L/4, etc. The smaller the wave length the greater the frequency. Since we assume that the string may be subdivided without limit, it would appear that beyond any given range of frequencies there would always be an infinite number of higher frequencies possible. No one frequency ought to be favored over any other, and we therefore expect that energy introduced at one frequency would soon excite higher and higher frequencies of oscillation. The infinite number of high-frequency modes of oscillation would absorb and emit infinite energy. All the energy of the universe in its random exchanges of radiation should continually shift to higher and higher frequencies, passing from red to orange to blue and beyond the violet to lethal X- and γ-rays. From this ultraviolet catastrophe we are in fact delivered, God having anticipated Planck and the theory of quanta in the design of the universe.

In the graph of Figure 15 we see schematically the classical assumptions which lead to the Rayleigh-Jeans law of frequency distribution. The number of modes of vibration is infinite at high frequencies, falling rapidly at lower frequencies. The energy which can be absorbed per mode of vibration is on the average the same (Figure 15a). Then the energy (intensity) for any given range of frequency ν_2–ν_1 is given by multiplying the average energy per mode times the number of modes.

The frequency-distribution curve which is observed in experiments is the shape of the familiar Maxwell-Boltzmann distribution curve (see Figure 17b). But this M-B curve represents the most probable division of the energy among the particles of a gas. Does the radiation reflected back and forth within the cavity somehow comprise a "gas" of particles of light? Without

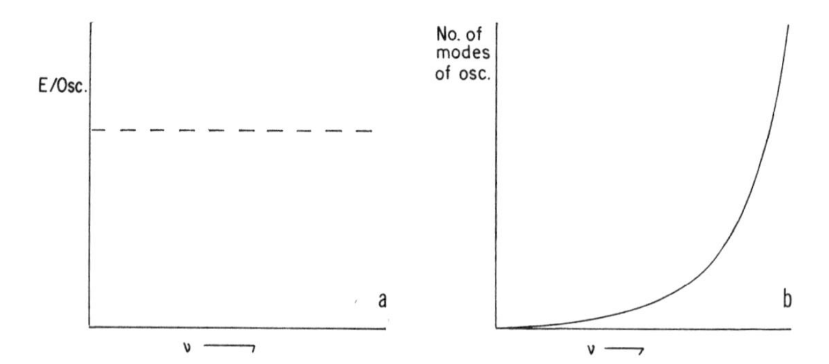

Figure 15. The Ultraviolet Catastrophe
a) The equipartition of energy; on the average, all oscillators absorb and emit equal quantities of energy regardless of frequency.
There are only a few ways of oscillating or vibrating at low frequencies. There are many ways of oscillating or vibrating at high frequencies. The longest wave, and therefore the lowest frequency, for an oscillator is fixed by the size of the oscillator. But there is no limit on how small a wave and therefore how high a frequency may be emitted or absorbed.
If all oscillators are to share equally in the energy, the very large numbers of oscillators of high frequencies will absorb and emit a larger proportion of the energy than the small numbers of oscillators of low frequencies. The emission-absorption curve of energy in any range of frequency would look like b, rising to infinity as the frequency increases.

going this far, Planck proposed that the atomic oscillators were capable of emitting or absorbing energy only in discrete parcels or units called *quanta.* The energy of each quantum was proportional to the frequency, which is to say that blue light is higher in energy than red—a result which at first seems unlikely, but which would immediately account for such facts as the penetration of matter by X-rays and the sunburn which requires ultraviolet light.

In Figure 16a the energy per quantum is shown as a function of the frequency; $E = h\nu$; ν is the frequency, h is the constant of proportionality (6.625×10^{-34} joule-seconds). In Figure 16b the number of ways of dividing a fixed amount of energy into quanta of each frequency is plotted as a function of frequency. Using a rough analogy, if we require that a class of students take an examination, and set the passing grade low enough, all of them will pass. If the passing grade is set higher, fewer of the students will pass. Very few, if any, will pass if the passing grade is set high enough.

Of one hundred low-grade problems, nearly all of our students will find one way or another to solve 60 percent of them. Of one hundred higher-grade problems, fewer students will find modes of solution for 60 percent. Of very high-grade problems, very few students will find ways of solving 60 percent. Similarly there will be many ways of absorbing and emitting quanta of low enough grade. There will be fewer ways to absorb or emit quanta of high grade. Instead of the energy being distributed equally among all possible frequencies, as Rayleigh's theory required, the random processes of reflection within the cavity would divide the energy among the quanta of various frequencies according to the number of ways of distributing quanta of such frequencies. If there are many ways of distributing quanta of the frequency of, say, red light, then the random exchanges within the cavity will insure that there are many oscillators excited to this frequency. If there are few ways of distributing quanta of a much higher frequency, then the likelihood of finding oscillators excited to this frequency is small.

The total energy given off within each small range of frequency will be the number of ways of distribution of the quanta times the energy ($h\nu$) of each quantum, or $E_v = nh\nu$ (n is the number). At highest frequencies the number of ways of distributing the energy in such large quanta becomes vanishingly small, so that the probability of exciting oscillations of highest frequencies is low, and relatively little of the energy is emitted in these frequencies. At the lowest frequencies the number of ways of distributing the energy becomes very great, but the energy per quantum becomes vanishingly small and the total energy emitted in this lowest frequency range is again small.

Planck was not so radical as to propose that the radiation within the cavity was itself in particles forming a kind of

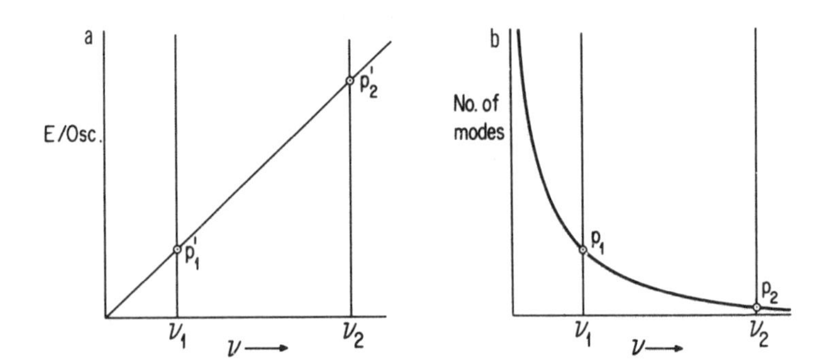

Figure 16. Planck's Hypothesis for the Black Body Emission Curve

a) The energy per oscillator is proportional to the frequency of oscillation; $E = h\nu$.

b) If a passing grade is set high enough, as at ν_2, few students (p_2) will pass. If the passing grade is moderate, ν_1, p_1 students will pass. The oscillators may absorb or emit energy only in whole units or quanta of size $h\nu$. The number of ways of dividing a fixed quantity of energy among oscillators of any one frequency depends upon the size of the quanta (and therefore the frequency ν). If the quanta are small, there are many ways of dividing up the energy. If the quanta are large, there are few.

At a high frequency, ν_2, there are few ways (p_2) of absorbing or emitting energy of quanta as large as p'_2. Then on the average, there should be $p_2 \times p'_2$ energy emitted by a black body of frequency ν_2. Since p_2 is close to 0, this product is small. At a moderate frequency, ν_1, there are p_1 ways of absorbing quanta of energy p'_1. Then the energy emitted on the average should be $p_1 p'_1$. Both are appreciable quantities and their product is therefore high. As ν approaches 0, the number of ways of absorbing or emitting energy rises sharply, but the energy per quantum approaches 0 and therefore the product pp' falls to nothing. The curve pp' vs. ν is the familiar skewed curve of the Maxwell-Boltzmann distribution (see Figure 17b).

"light" gas, but instead proposed that the atomic oscillators themselves were restricted to emitting and absorbing energy in discrete packets. But five years later, in the second of his 1905 papers, Einstein extended the concept of particle to light itself. The problem arose in the attempt to account for the photoelectric effect. Hertz first, then, more explicitly, Philipp Lenard (1862–1947) had observed and recorded that electrons were given off when metal plates were irradiated with ultraviolet light. There were two experimental laws of the *photoelectric effect,* as it was called. First, the *voltage* of the electron discharge (or the energy imparted to each of the electrons) depended on the frequency—that is, color—of the light above a certain minimum or cut-off frequency. This voltage had nothing to do with the intensity—which is to say, strength or brightness—of the light. In the second place, it was the *current* of the discharge (or the number of electrons per second) which was determined by the intensity of the light.

Einstein's simple but radical explanation was that the discharge was effected by the bombardment of the metal by corpuscles (quanta) of light which he called *photons.* Following Planck, the intensity of the light would be related to the number of quanta, while the energy of the photons would be proportional to their frequency. In dislodging an electron from a metal the energy of a single photon $h\nu$ would be used for the work (W) of separating the photon from the metal, and beyond that, to impart to the electron kinetic energy of $\frac{1}{2}mv^2$; $h\nu = W + \frac{1}{2} mv^2$.

This theoretical treatment covered the empirical laws of Lenard exactly, but Lenard was hardly pleased. He came to believe that all of theoretical physics was a trick designed to steal the credit of discovery from the experimentalists. He was in the unhappy position of having his painfully won experimental results explained for him by brilliant boys. Hungarian by birth, Lenard achieved a high position in the Prussian university hierarchy. He became a fanatic German nationalist. One has only to view the portrait of Einstein in 1912—the curly black hair, the delicate features—to imagine the fury that Einstein's wit and grace engendered in proto-Nazi minds. Ignoring the fact that Hertz, his idol, had been of Jewish origin and as devoted to the theoretical side of physics as the experimental side in which he made his reputation, Lenard, with Stark, became the leader of a "German science" movement, an anti-intellectual, anti-Semitic movement which reached its nadir in the

Nazi era. This was the reintroduction of political passion to science, the conflict of man the political animal—whose first concern is obtaining and keeping power—with man the husbandman, whose concern it is to mold the material universe by the medium of thought.

The extension of the quantum hypothesis of Planck to the photoelectric effect which Einstein presented modestly as "a heuristic point of view" was followed in the next few years by the further extension of the quantum hypothesis to the puzzling problem of heat capacities (the amount of heat absorbed or released by a substance in changing temperature). In 1901 Gibbs had found the kinetic-molecular theory unable to account for the heat capacities of gases, and he had written that ". . . it seems hardly possible to frame a dynamic theory of molecular action which shall embrace the phenomena of thermodynamics, of radiation, and of the electrical manifestations which accompany the union of atoms."

The heat capacities of solids at ordinary temperatures obey the law of Dulong and Petit, that is, energy in the form of heat is divided equally on the average among the modes of vibration of the solid.* But if agreement with the classical theorem of the equipartition of energy was in general good, it was hopelessly inadequate for light elements or for any elements at low temperatures (Figure 17a). At temperatures in the vicinity of the absolute 0, all heat capacities become extremely low—that is, very small amounts of heat will produce major changes in temperature.

Einstein was able to account for this in 1906 with a further

---

* The N atoms of a mole of an elemental solid (such as metallic copper) are free to vibrate in any direction; this results in components of vibration in the x, y, and z Cartesian coordinate directions. Not only energy of motion ($\frac{1}{2}mv^2$) is taken up by the vibrating atoms, but since there are bonding forces directing the atoms to their mean positions, the displacement of an atom takes up potential energy. Displacements are resolved into components along the same three coordinate directions as the motion. This gives 3 + 3 or 6 ways of taking up energy per atomic vibrator (called degrees of freedom). Classical theory would say that these are on the average equal. If a unit of heat (calory) is the energy absorbed per degree Centigrade of temperature rise, for each degree of freedom, then the total heat capacity of elemental solids should be 6 calories per mole per degree.

extension of the quantum theory, now to include heat, and with it thermodynamics. He assumed that the thermal vibrations of the atoms of a solid were all of the same frequency ν, and that energy would be apportioned among them in the form of vibration quanta (phonons) of energy $h\nu$. No atom would be excited except by the absorption of at least one phonon. If the total amount of energy to be communicated was small, it could be divided into only a few phonons of energy $h\nu$. Only a fraction of the oscillators could begin to vibrate and the capacity for heat should therefore be low. At higher temperatures, the number of subdivisions into phonons $h\nu$ would be greater. The probability that an oscillator will be excited rises. At still higher temperatures, the energy available per atomic oscillator is overwhelmingly greater than $h\nu$ and the oscillators can then on the average divide the energy equally. The law of Dulong and Petit will then obtain.

Modifications of the theory, by Peter Debye (1884–1967) in 1912, and by others later on, involved more complicated spectra of atomic vibrations than Einstein's first approximating assumption that all vibrations were of the same simple frequency. We know now that all quantum ideas on the specific heats give results in good agreement with observation, regardless of the particular nature of the assumptions about the frequency, so that it is the single requirement of discreteness or corpuscularity in the energy exchange which here for molecular thermodynamics, as before for radiation, will account for the phenomena which classical theory would have denied existence.

With these further successes of the quantum hypothesis, it passed from the status of a brilliant ad hoc assumption accounting for the peculiarities of the minor phenomenon of the black body radiation—or even a heuristic explanation of the photoelectric effect—to the single most fertile idea of the century and the foundation of all that prodigious burst of scientific activity which we see today. By 1906 it had already provided that dynamic theory which had seemed hardly possible to Gibbs in 1901, embracing the phenomena both of radiation and of thermodynamics. In 1913 Bohr was to extend the theory to cover the third of Gibbs' requirements, the electrical phenom-

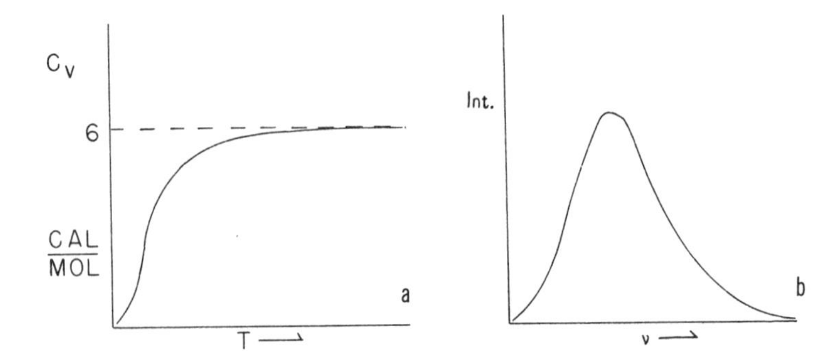

Figure 17. The Heat Capacity of Solids
a) If all the particles of a solid vibrate so as to share equally in the energy, the heat capacity should be six calories per mole-degree. It is this value at ordinary temperatures but decreases to 0 at the absolute 0, showing that the particles do not divide up the energy equally at very low temperatures.
b) The black body radiation distribution curve: The intensity vs. frequency curve is of the same form as the curves for the proportions of particles traveling at different velocities. If the total area under the curve starting from the origin is plotted against the frequency, a curve of the form a), the heat capacity curve results. This can be explained on the hypothesis that at low temperatures there are not enough phonons of energy $h\nu$ to set many of the particles vibrating. At higher temperatures, larger numbers of phonons may activate most of the particles. Finally, at still higher temperatures, all the particles vibrate and, on the average, share equal proportions of phonons as in the classical theory. At these temperatures the heat capacity reaches the Dulong and Petit value of six calories per mole-degree.

ena accompanying the union of atoms. This extension, more than any of the others except possibly the theory of the photoelectric effect, was the direct result of experiment.

The period was one of explosive fertility. The disturbing results of the Michelson-Morley experiments which led to the theory of relativity were by no means the most revolutionary developments to come from the laboratory. The integrity of the atom itself was breached, a research which originated with Faraday and his attempts to understand the fine structure of matter through experiments on conductivity. Just as he had measured the conductivity of solutions and in this way developed the subject of electrolysis, Faraday with his infallible sense of the right experiment had tried to measure the conductivity of gases. If a high potential difference is placed between electrodes in an evacuated tube, current will flow accompanied by various luminescent and phosphorescent glows. A "negative glow" proceeds from the cathode (negative electrode) toward the anode.

J. W. Hittorf (1824–1914) fashioned the cathode into the form of a point. "There then appears around this a beautiful blue hemisphere of glow . . . whose radius increases very fast as the density [of the air] diminishes." This glow "brings the surface of the glass to fluorescence," Hittorf reported in 1869. Because intervening objects cast sharp shadows, he spoke of "rectilinear paths or rays of glow," which became known as the cathode rays. Crookes (1832–1919) referred to these as *molecular rays,* saying that they radiated in straight lines from the negative pole, "casting strong and sharply-defined shadows of anything in their path." He mounted a small pinwheel version of his radiometer and demonstrated that the cathode rays set the wheel to rotating. The wheel is matter, and to move matter is to impart momentum, or mass times velocity, to it. According to a well-known law of mechanics, momentum cannot be created or destroyed but can only be transferred by impact. The cathode rays transfer momentum and therefore must indeed be rays of particles with mass and velocity. The "molecular" impact, as Crookes called it, heated objects to incandescence on bombardment and the "molecules" were charged, as he could show by their deflection in magnetic fields.

Carrying this a step further, J. J. Thomson (1856–1940), by comparing the deflection of the beams in magnetic and electric fields,* obtained a rough measurement for the ratio of the charge of the cathode particle to its mass.

* J. B. Perrin, by passing the cathode rays into a closed conducting cylinder connected to an electroscope, established the negative charge of the particles.

> It was sufficient however to prove that e/m for the cathode ray particles was of the order 10^{-7}, whereas the smallest value hitherto found was 10^{-4} for the atom [ion] of hydrogen in electrolysis. So that if e were the same as the charge of electricity carried by an atom of hydrogen . . . m, the mass of the cathode ray particle could not be greater than one thousandth part of the mass of an atom of hydrogen, the smallest mass hitherto recognized.

By this discovery of new particles with masses a thousand times smaller than the least of the atoms, the concept of atomism, of hard, round, unchanging and unchangeable, indivisible monads, was overthrown in the very period of its final establishment.

The gas tube researches and the techniques for studying the rays and the fluorescence of the tubes could be said to have originated with the seventeenth-century experimenters of the Accademia del Cimento in Florence, who had observed electrical effects in the gases of flames. These experiments had been revived when Davy deflected the carbon arc with a magnet, and by Faraday's gas tube studies. Geissler had invented the mercury vacuum pump in 1855, enabling the production of vacuums below 1 mm. pressure used by Hittorf and all later observers.

Eugen Goldstein (1850–1930), observing a yellow glow in the vicinity of the cathode, bored "canals" through the cathode and found streaming back from every opening in the cathode ". . . a bright slightly divergent yellow beam of rays." Goldstein was unable to deflect these canal rays with magnetic fields, but Wilhelm Wien (1864–1928) succeeded in establishing that they were positively charged particles with a charge-to-mass ratio of about 10^{-3}, which is about right for ions of gas. They traveled with a velocity of 3.6×10^7 cm./sec., 1/1000th the velocity of light. Thomson's value for the velocity of the cathode ray particles was 1/10th the velocity of light.

Wien's publication was in 1898, twelve years after Goldstein's observations and three years after the discovery of X-rays. Wilhelm Konrad Roentgen (1845–1923) accidentally observed fluorescence produced near a cathode ray tube which was completely covered with black cardboard. "This fluorescence is vis-

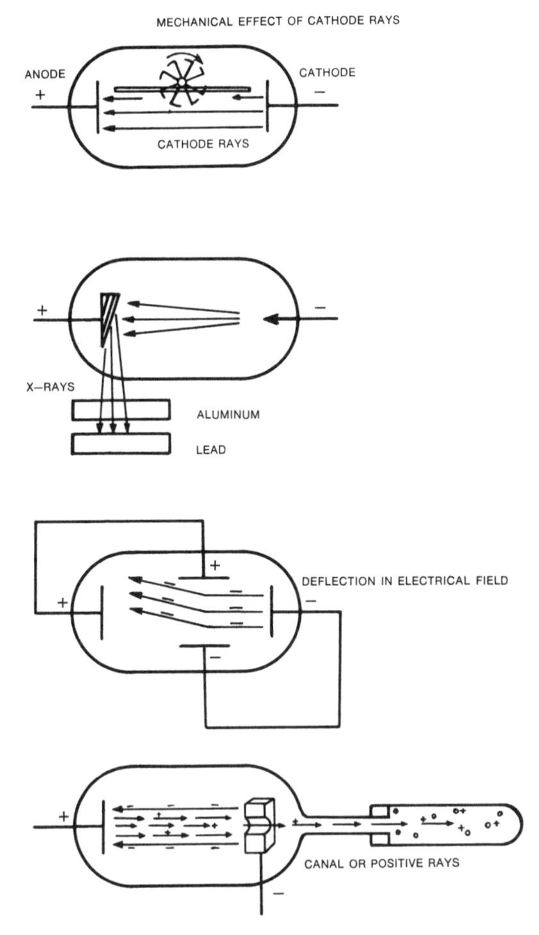

Figure 18. Cathode Rays

ible even when the paper screen is at a distance of two meters from the apparatus," he reported. The source was a new, invisible, and highly penetrative radiation emitted from matter on being struck by cathode rays.

Here too Philipp Lenard had missed great discoveries. With Hertz, he had reported on the penetration of aluminum by what he took to be the cathode rays. He argued from this that the cathode rays had to be hertzian or electromagnetic waves and could not be streams of material particles. Roentgen used Lenard's arguments to justify his conclusion that the X-rays were electromagnetic waves. Finding them still strong at a distance of over two meters from the apparatus, compared to perhaps a hundredth part of this for the cathode rays, ". . . I conjecture that in Lenard's experiments it was the X-rays and not the cathode rays, which had passed unchanged through the aluminum window of his apparatus . . ."

A certain drama attended these discoveries, enhanced if anything by Lenard's claims to priority. On first presenting his paper on the properties of the new rays, Roentgen illustrated his discovery by preparing a radiograph of the bones of the hand. The German Physical Society immediately voted to name the new radiation "Roentgen Rays" in honor of its discoverer.

Using the new concept of invisible penetrating radiation and the techniques so vividly demonstrated by Roentgen, Henri Becquerel (1852–1908) explored the phosphorescence of uranium salts exposed to sunlight. He found that the salts gave off radiation which would penetrate thicknesses of black paper impenetrable to the sunlight; then, systematically and elegantly refining his experiments, he demonstrated that the radiation did not require sunlight for its excitation and that the source was the element uranium. It was the discovery of radioactivity.

In 1897, two years after Becquerel's finding, Marie Curie, measuring the radioactivity of various minerals containing uranium, obtained anomalously high readings. These could be explained only by the assumption that the minerals contained small amounts of a new element, much more highly radioactive than uranium. To isolate the new element, Marie and Pierre Curie used its radioactivity, measured as the rate at which the radiation in each successive chemical fraction would discharge an electroscope. The new element, of course, was radium, and with it an entire series of related elements, the newest members of which are still being discovered.

Becquerel had discovered an invisible radiation which he first thought to be X-rays.

> . . . as on those days the sun only showed itself intermittently I kept my arrangements all prepared and put back the holders in the dark in the drawer of the case, and left in place the crusts of uranium salt. Since the sun did not show itself again for several days I developed the photographic plates . . . expecting to find the images very feeble. The silhouettes appeared on the contrary with great intensity.

These rays were immediately tested by Ernest Rutherford (1871–1937), an enormous redheaded New Zealander who had joined J. J. Thomson at Cambridge. Thomson, unable to pursue research at this level with undergraduates, had invented the postgraduate research group, a kind of scientific analogue to the graduate schools in the liberal arts then flourishing in the German universities. In studying the ionization of the air produced by the radiation Rutherford found that there were different kinds of radiation coming from the uranium; one kind, which he called α-rays, was easily absorbed, while a second kind, the β-rays, was hundreds of times as penetrating. Becquerel and others reported at first that the α-rays were not deflected in a magnetic field but that the β-rays were. Soon afterward the Curies showed that the β-rays were negatively charged. Becquerel then succeeded in deflecting them with an electrostatic field, finding values for the velocity and e/m ratio which identified them with cathode rays.

Rutherford found values for the velocity and e/m ratio of the α-rays to be similar to those of the canal rays. He later succeeded in identifying them as streams of ionized helium atoms. A third stream of radiation, undeflected by either electrostatic or magnetic fields, was discovered by Villard in 1900 and called γ-radiation. Highly penetrative, this radiation was eventually identified as ultra X-radiation, which is to say, hertzian electromagnetic waves like light but of much higher frequency than any others known.

In 1899 Rutherford accepted a chair at McGill University, in Montreal, where he was joined by Frederick Soddy (1877–1956). Rutherford and Soddy published a series of papers proposing a general theory of radioactivity. The radioactive elements, they asserted, spontaneously transmute into other radioactive elements, these transformations taking place at rates characteristic of each element. The energy emitted in the form of rays of charged particles had its source within the atom, and was therefore distinct from the energy of ordinary chemical reaction. The loss of charged particles by one atom transforming into another was itself the transmutation. In modern terms radium atoms transmute to radon atoms plus α-particles. Radon atoms in turn transmute to polonium atoms plus α-particles. The transmutations continue until a stable daughter element with the chemical properties of lead is reached. For the series beginning with uranium there are fifteen separate transmutations ending in stable lead.

In these researches the individual species were identified by their radioactive behavior, following the Curies' original procedures. Rutherford and Soddy had proposed that radioactive decay occurred at a constant rate, the number of atoms transmuting at any time being a definite fraction of the number of atoms present.* This fraction, the *decay constant* λ, is the identifying characteristic of each radioactive species. Literally dozens of new radioactive elements were being uncovered as the study of the decay products proceeded. In some instances—notably in the case of Boltwood's ionium, which was the direct parent of radium, and the radio-thorium of O. Hahn, and also in that of ordinary thorium—elements with markedly distinct radioactivities were found to be chemically indistinguishable, even yielding indistinguishable optical spectra.

This concept of elements that were physically distinct but chemically identical was established when J. J. Thomson took up the study of the canal rays discovered by Goldstein, which he called positive rays. Unlike those of the cathode rays, which were all alike whatever their source, the charge-to-mass ratios of the positive rays depended on the gases in the tubes. By passing the canal rays between magnetic and electrostatic deflecting plates at right angles to each other, the particles would be deflected from a straight-line course according to their mo-

* $n = n_0 e^{-\lambda t}$; n_0 is the initial weight of the radioactive species present, n is the weight at the time t, e is the number 2.718 . . . , and λ is the decay constant for the species.

mentum (mass times velocity) and their electric charge. All the particles with like mass and charge would be spread along a parabola according to their velocities. Particles with different charges and masses would fall along distinct parabolas, one for every species of gas in the tube. The relative weights of the particles could be derived from the parabolas. In 1913, when Thomson delivered the Bakerian Lecture before the Royal Society, he discussed his finding of two parabolas for neon gas, one giving an atomic weight of 20, the other of 22. There were two distinct kinds of neon gas. It was now clear that the elements of the periodic table were really groups of what Soddy termed *isotopes,* physically distinct atomic species with identical chemical properties.

In radioactive transmutation the emission of an α-particle (an ion of helium, atomic weight 4, atomic number 2) reduces the atomic weight of the atom by 4 and lowers the atomic number by 2. The uranium series in modern notation begins with ${}_{92}U^{238}$; the subscript stands for the atomic number or position on the periodic chart, the superscript represents the atomic weight. With the emission of a single α-particle, the uranium atom transmutes to an atom of weight $238-4 = 234$ and number $92-2 = 90$, (${}_{90}Th^{234}$). The ninetieth element on the periodic table is thorium. Thorium 234 spontaneously emits a single β-particle. The emission of a β-particle produces no significant change in the atomic weight (1/1800 of unit atomic weight), but since the β-particle is an electron or unit negative charge, the chemical behavior of the atom of ${}_{90}Th^{234}$ is shifted up one place in the periodic chart to ${}_{91}Pa^{234}$ (protactinium). With the emission of another β-particle, the protactinium atom transmutes back to uranium, but this time it is ${}_{92}U^{234}$ instead of ${}_{92}U^{238}$, with which the series began.*

* The complete uranium series is:

$${}_{92}U^{238} \xrightarrow{\alpha} {}_{90}Th^{234} \xrightarrow{\beta,\,\gamma} {}_{91}Pa^{234} \xrightarrow{\beta,\,\gamma} {}_{92}U^{234} \xrightarrow{\alpha} {}_{90}Th^{230} \xrightarrow{\alpha,\,\gamma}$$

$${}_{88}Ra^{226} \xrightarrow{\alpha,\,\gamma} {}_{86}Rn^{222} \xrightarrow{\alpha} {}_{84}Po^{218} \xrightarrow{\alpha} {}_{82}Pb^{214} \xrightarrow{\beta,\,\gamma} {}_{83}Bi^{214} \xrightarrow{\beta,\,\gamma}$$

$${}_{84}Po^{214} \xrightarrow{\alpha} {}_{82}Pb^{210} \xrightarrow{\beta,\,\gamma} {}_{83}Bi^{210} \xrightarrow{\beta} {}_{84}Po^{210} \xrightarrow{\alpha,\,\gamma} {}_{82}Pb^{206}\text{(Stable)}.$$

We live today with many of the fruits of these researches but especially the fear of atomic warfare, glancing uneasily at the headlines on the newsstand, as the last of the Sung Emperors must have paused on his stroll through the imperial gardens. For the distant rumble may be only the approach of a spring shower, or it may be the sound of the Mongol cannon, and General Subotai stands finally on the southern bank of the Yellow River.

14

Crystallographic Atomism

WHENEVER IT BEGINS TO SNOW, THE FIRST BITS OF SNOW SHOW THE FORM OF SIX-RAYED STARS. THERE MUST BE AN EFFECTIVE CAUSE FOR THIS.

—KEPLER, 1611

In the twentieth century our generation has seen the atomic bomb. We have gone through the revolution in solid state physics; we have succeeded in placing men on the moon and we have transplanted human hearts, yet none of these were the most arresting developments in the science of our times. The most dramatic field of discovery in our generation and our claim to parity with the age of Kepler and Galileo is the field of the structural chemistry of living things. It is here that we have revived mechanism and dramatically realized the hopes of the seventeenth century. It is here that otherwise sober and cautious men speak of their work in phrases such as "the secret of life."

Immediately after the Second World War great advances in automation, electronics, and the development of computing machines placed the whole subject of the X-ray diffraction of organic compounds on a new plane. It was useful and important for complex and difficult structures such as the antibiotic penicillin to be resolved, but significant results for the first time were being obtained for the proteins and the nucleic acids. For the first time it was possible to seriously anticipate a rigorous theory of life and heredity based on physical-chemical law. Plausible mechanisms for the transmission of hereditary characteristics, for animal motion, for photosynthesis, for growth and for disease were being proposed on the basis of detailed analyses of the numbers and configurations and motions of atoms. The extension of man's control of nature to the determination of the direction of his own evolution is now within the realm of possibility.

It was René-Just Haüy (1743–1822), the modest son of a poor cloth worker, whose accidental destruction of a fine group of calcite crystals revived the study of geometric crystallography in 1781. Coming late to the study of mineralogy from botany, Haüy was struck by the observation that crystals even of the same salt might occur as cubes or prisms, as needles or plates without the slightest change of composition, while the much more complicated forms of the leaves and flowers of plants of the same species were always the same. Visiting a friend, Haüy dropped a specimen of prismatic calcite. On stooping to pick up the fragments he observed that these were bounded by perfectly smooth faces intersecting at the characteristic angles of a rhombohedron and duplicating exactly crystals of Iceland spar.

Returning to his home, Haüy smashed crystal after crystal of calcite, always obtaining rhombohedral cleavage fragments, although the original crystals might have been tablets or hexagonal prisms or other complicated forms. He at once concluded that the *molecule* of calcite must be in the shape of a rhombohedron. Similarly the molecules of fluorspar were octahedrally shaped, those of pyrite were cubes, and so on.* It was a revival of homoiomerism (see page 27) but with a significant change. Haüy's homoiomeres were ultimately molecules. From the shape of the molecules he accounted for the form of crystals. His contemporaries, indignant at having themselves failed to make so important a discovery, called him "cristalloclast," at once denying the truth of his hypothesis and claiming for themselves priority in its discovery.

Since the time of Hooke and Steno, the precise measurement

* Pyrite has no cleavage form, but Haüy inferred the *nucleus* shape from the crystal form.

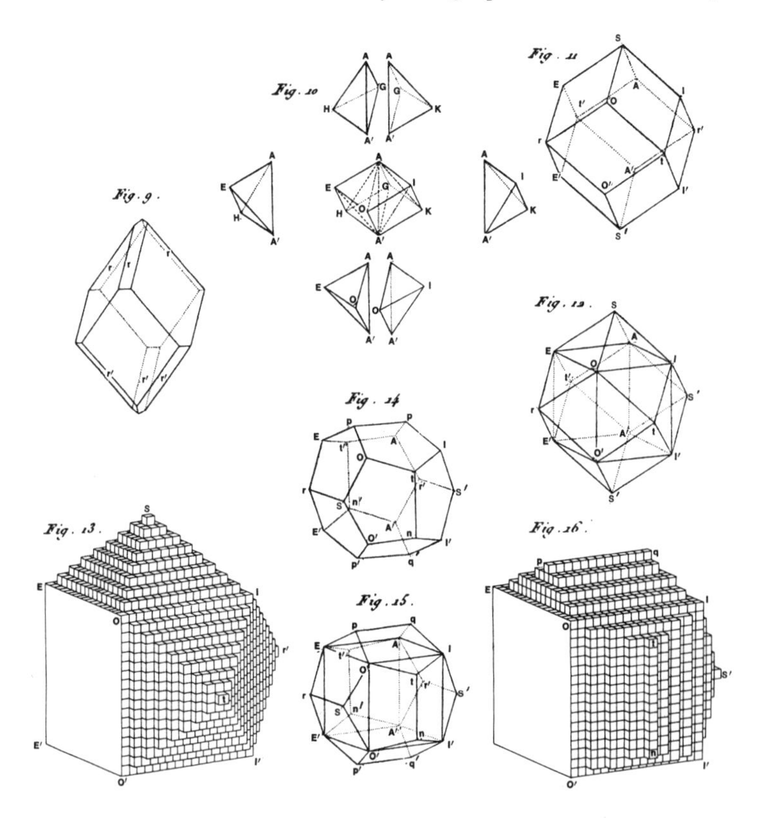

Figure 19. Haüy's Molecules, 1801
The shapes of the molecules which would pack together in orderly fashion to make the plane faces of the crystal were the shapes of the cleavage fragments.

of the angles between the plane-bounding faces of crystals had become a principal task of mineralogists, who used the data for identification. With Haüy's idea that the external plane faces of crystals were the boundaries of an internal stacking of block-shaped molecules, it became possible to refer all the measurements of a mineral species to the proportions of a fundamental *nucleus*. Considering a face of a natural crystal of diamond in the shape of an octahedron (see Figure 1, page 18) and referring all angular measurements to three equal axes intersecting at right angles, the number of molecules from the center to where the plane intercepts one axis is the same as the number of molecules to the intercepts of the other axes. This ratio of 1:1:1 will be constant for large as well as small crystals. Haüy ascribed it to the fundamental *molécule intégrante*. Other, more complicated facets might occur on the diamond, in which case the intercept on an axis might be twice that for the octahedral facet, giving a ratio of 2:1:1, or three times, or any rational number since it must always comprise a whole number of molecules. This, the law of rational intercepts, is the geometric counterpart of Dalton's later law of integral combining proportions. It is consistent with the hypothesis that the matter of the crystal is disposed in discrete units—the shaped molecules of Haüy—but even more important, it requires the arrangement of these units of structure to be uniform, orderly, and repetitive. The structure of crystalline matter is *periodic*.

The student of crystallography is rapidly aware of the possibility of geometric classification. For example, the ordinary salt crystal and its cleavage fragments are cubic. The snow crystal is patently hexagonal. The fragments of Iceland spar were rhombic. Auguste Bravais (1811–1863) altered the Haüy idea subtly by proposing to substitute lattices of ideal points placed at or between the corners of the nuclear molecules in Haüy's periodic structures. Bravais' lattices were geometric constructions to which the physical units of structure, such as atoms or ions or molecular groups, could be referred.

By systematically exploring the possibilities, Bravais discovered that there were fourteen different ways of distributing equivalent points periodically through space. The simplest plane net is one of points spaced in two parallel sets of rows. If the two rows intersect at right angles, there are two possible nets, one of squares, the other of rectangles. A third possibility would be the net of equilateral triangles generated if both sets of rows intersect at a 60-degree angle. The actual unit of structure, atom or molecule, might itself be simple or it might be highly complicated.

Bravais' representation is essentially *aniomeric* (see page

28), since the geometry of the crystal is derived from the way the units of structure are arranged in space, rather than, as in Haüy's hypothesis, from their shapes. Each element of the complex formula-molecule of a crystal could be referred to the points of a separate lattice. For example, CsI crystallizes in cubes. Imagine the cesium ions located at the corners of a cubic lattice and the iodine ions located at the corners of another cubic lattice of the same size. The array of iodine interpenetrates the array of cesium, so that each iodine is at the center of a cube of 8 cesium and each cesium is at the center of a cube of 8 iodine. Given a complex silicate such as feldspar, $KAlSi_3O_8$, let each separate unit of the formula fit the points of a lattice—a three-dimensional array of potassium ions, another geometrically identical array of aluminum ions, three separate congruent arrays of silicon ions, and eight of oxygen—all interpenetrant and together composing the structure.

The problem for Bravais, as for anyone else speculating that the atoms or ions in solids retained their identity in a discrete periodic structure, was the complete absence of any way of experimenting on atomic structures. It was not possible in 1848 to analyze objects of the size of atoms. Bravais proposed to arrive at the structures of crystals by assuming that their flat surfaces would correspond to planes through the lattice with the most points. Thus crystals, such as garnet, that commonly occur in rhombic-sided dodecahedra he referred to a cubic lattice with a point at the center of each cube, because dodecahedral planes through such a lattice have the highest density of points. Although they were plausible, there was no way of proving that his arrangements were right.

The modern computer technology which was designed to locate the positions of the atoms within the crystal from X-ray data confirms Bravais' idea. The X-ray crystallographer assumes that the planes through the crystal with the most electrons will give the strongest X-ray signals. These planes contain the most and heaviest atoms. From the intensities of the diffracted X-rays an elaborate technique (Fourier synthesis) is used to calculate a map of the electron densities in a slice through the crystal. If as Bravais suggested the most prominent surface planes of the crystal are also parallel to the planes with the greatest number of atoms, the same analysis should give the same map. Using statistical measures of different surface planes of crystals of anglesite (lead sulfate) in place of X-ray data, this method yields an electron density map clearly showing where the lead atoms (with the highest number of electrons) are located (Figure 20).*

The study of the symmetry of crystals (the study of their

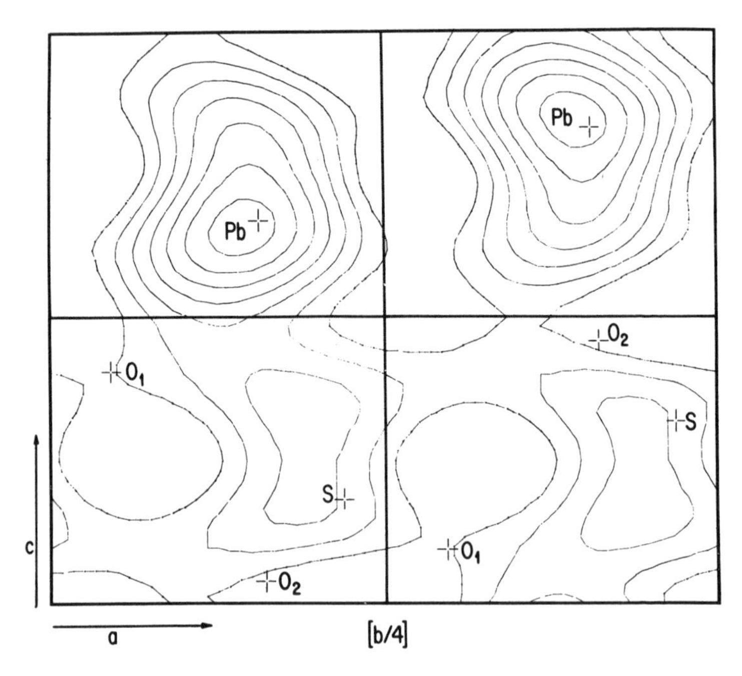

Figure 20. Map of the Electron Concentration in the Mineral Anglesite
A single Haüy *molécule intégrante* or *unit cell* of the mineral anglesite, $PbSO_4$; the concentration of electrons in a slice ¼ of the distance from the base of the cell is represented by contour lines of equal electron density. Plus signs mark the locations of the centers of the atoms located in the plane of the slice, known by other methods of analysis. The map was made from statistical studies of the crystal form by the method of Fourier synthesis.

* Schneer, 1968.

congruences) was exhaustively pursued in the half-century after Bravais. Certain operations, for example rotations about axes, reflections across planes, or inversion through a point, leave the configuration of the crystal unchanged (Figure 21). The

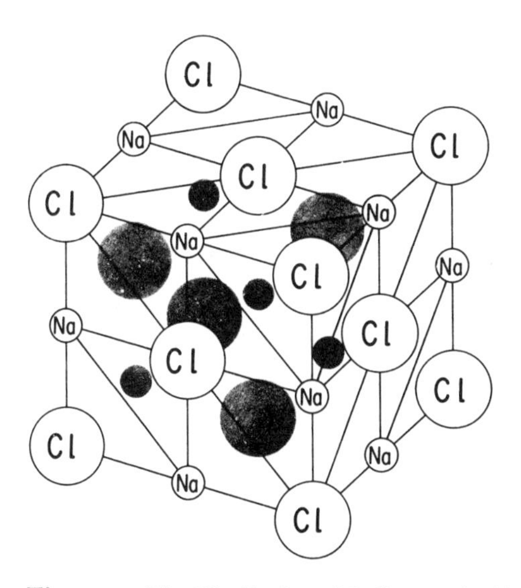

Figure 21. The Distribution of Sodium and Chlorine Atoms in Rock Salt

The diagram shows some of the symmetry of the structure. Considering that the arrangement extends repetitively in all directions, the surroundings of one chlorine atom (Cl) are the same as the surroundings of any other. Therefore, a shift of the whole model in the direction of and for the distance between any two chlorine atoms would leave the configuration of space unchanged. Such translations are examples of symmetry operations. A rotation of 120° about the diagonal of the cube would also bring Cl's into coincidence with Cl's, and Na's into coincidence with Na's, so that the total configuration would again be the same. This is another example of a symmetry operation. Along the diagonals of the cube, the structure is one of alternating planes of sodium and chlorine atoms. Along the edges of the cube, planes are identical checkerboard patterns of alternate sodium and chlorine atoms. The interaction of waves and the atomic structure depends upon the direction of the waves relative to the structure.

sum of these operations, which is to say of the geometric regularities of the crystal, is its symmetry. Crystal forms could be classified on the basis of their symmetry. Some, like the human figure, are bilaterally symmetrical. Some, like the clover leaf, are trigonally symmetrical. Considering that the fundamental property of crystalline matter is periodicity and any symmetry must be compatible with the repetitive symmetries of Bravais' lattices, it was possible to show that there were only five kinds of pure rotation possible. The axes to which the geometry of a crystal could be referred could exhibit unique, twofold, threefold, fourfold, or sixfold rotational symmetry but not pentagonal or greater than hexagonal symmetry. Crystal forms could show rotary symmetry and reflexive symmetry* in thirty-two distinct combinations or *point groups*. Every crystal form can be referred to one or another of these groups, called point groups because their operations of symmetry leave a single point unmoved.

These symmetry limitations of the external form also apply to the configurations represented by the points of the fourteen Bravais lattices (which are simply identical points of a periodic configuration, like the points of repetition of the complex motif of a wallpaper pattern). Within these restrictions, the combinations that are consistent and distinct fall into 230 *space groups* derived independently by the Russian crystallographer E. S. Federov (1853–1919), the German mathematician A. M. Schoenflies (1853–1928), and the London mineralogist William Barlow (1845–1934).

In 1883 Barlow wrote:

> . . . in the atom-grouping which modern chemistry reveals to us the several atoms occupy distinct portions of space and do not lose their individuality. . . . this conclusion is in harmony with, and indeed to some extent explains, the symmetrical forms of crystals, and the argument may therefore . . . be considered an extension of internal symmetry derived from the phenomena of cleavage.

---

* Essentially the symmetry of the object and its undistorted image either through a point lens or in a mirror.

At that time he was apparently unaware of the work of Bravais, since he wrote that "... there are but *five* (very symmetrical arrangements of points or particles in space) . . ." possible. Linus Pauling writes that it was apparently the sharp criticism of Barlow's limited views on symmetry by the German mineralogist Leonhardt Sohncke that spurred Barlow to undertake the laborious derivation of the 230 space groups.

By this kind of reasoning G. H. Tammann, W. J. Sollas, F. M. Jaeger, but especially Barlow and William Pope (1870–1939), a Manchester chemist, arrived at reasonable atomic structures for the simpler crystalline compounds, but of course with no way of verifying their ideas until X-ray diffraction was introduced in 1911. The discovery came about in the attempt to explain the new X-rays.

The German physicists were convinced that Roentgen's X-rays were electromagnetic wave phenomena like light. Arnold Sommerfeld (1868–1951), who had replaced Boltzmann at Munich, discussed with his students the attempts at diffraction of X-rays through a crude slit by Walther and Pohl at Hamburg. English physicists, notably W. H. Bragg (1862–1942) at Leeds, were trying to prove that the rays were corpuscular. P. P. Ewald (1888–   ), a student of Sommerfeld's, was walking back from the university with Max von Laue (1879–1960), who was then a *privat-docent** and did not yet have the prefix von to his name. Ewald had assumed that the optical effects in crystals which he was studying were due to the action of atomic oscillators in a lattice array.

"What is the distance between the resonators?" he recalls von Laue inquiring as they crossed the street and entered a park called the *Englische Garten*. On learning that the lattice spacing within crystals was of about the same size as the

---

* Roughly equivalent to a lecturer, a *privat-docent* in the old German universities originally had an appointment, the privilege of working, and the job of teaching, but, at least in the old days, no pay. *Privat-docents* could charge admission to their lectures and stand at the door to collect the fees, delaying the start of the lecture until the last possible minute to get the fees of latecomers. Impecunious students would try to get in after the lecture had begun. In his middle or later years, if he survived, the faithful *privat-docent* might receive an appointment as professor.

hypothetical wave length of X-rays which Sommerfeld had obtained from the Hamburg experiments, it occurred to von Laue that this would provide a definitive test of the wave hypothesis. If X-rays were waves, they should be diffracted by the natural three-dimensional grating of a crystal. The experiment was discussed by the young physicists who used to linger over coffee and cakes at the Café Lutz in the *Hofgarten,* scribbling equations all over the white marble table tops. Walter Friedrich, (1883–) and Paul Knipping (1883–1935), students of Roentgen's, undertook the experiment, which failed on the first try.

Because of an erroneous analogy with optical diffraction they had placed the photographic plate *between* the X-ray source and the crystal. They also used the first crystal which came to hand, a crystal of copper sulfate, which has the lowest possible symmetry. But on their second exposure they obtained a plate with diffracted spots. It was positive proof of the discreteness and the periodicity of crystalline matter, as well as of the wave nature of the roentgen radiation.

The copper sulfate was replaced by a crystal of sphalerite, a highly symmetric zinc sulfide. Von Laue showed that the pattern of diffraction spots not only yielded the cubic symmetry of the sphalerite but could be used to derive the dimensions of the structure (such as the length of the cubic Haüy *molécule intégrante*). But many different arrangements are possible for zinc and sulphur which still preserve the ratio of one zinc to one sulphur. The symmetry, even the lattice type, while narrowing down the possibilities, does not uniquely prescribe the arrangement of zinc and sulphur atoms which is the structure.

When W. H. Bragg learned of the von Laue experiments in the summer of 1912, he discussed them with his son, W. Lawrence Bragg (1890–), then a student at Cambridge. The Braggs at first tried to explain von Laue's, Friedrich's, and Knipping's remarkably clear ZnS plates as somehow caused by "'X-ray corpuscles' shooting down the avenues between the rows of atoms in the crystal." Back at Cambridge, and at first considering the X-rays as particles being mechanically sorted by tunnels through the crystal grating, W. L. Bragg hit upon the idea of explaining the spots on the Laue diagram as due to the reflection of X-rays from "sheets of atoms in the crystal as if these sheets were mirrors." The two Braggs showed that on this basis the X-rays reflected from successive parallel planes would interfere constructively when their wave length λ, angle of incidence (and reflection) θ, and the spacing between planes, d, satisfied the equation

$$\lambda = 2d \sin \theta.$$

Next W. L. Bragg undertook to explain why some of the reflections were of greater intensity than others, by the differences in the reflecting planes. He tried Barlow's model for the close-packing of spheres and explained the anomalies in the von Laue ZnS diagram. Bragg Senior had meanwhile determined that the X-rays included strong monochromatic radiation characteristic of the element used for the anti-cathode of the X-ray tube. He designed a spectrometer that made possible the direct comparison of the diffracted intensities. Comparing X-ray patterns for NaCl, KCl, KBr, and KI, "by noticing what differences were caused in the photograph by the substitution of heavier for lighter atoms in the crystal," the exact positions of each of the atoms in the Haüy *molécule intégrante* (now transformed into the *unit cell*), were worked out.*

Friedrich had insisted that little more than the internal symmetry could be derived from X-ray examination. Following

---

* In the Pope and Barlow model for rock salt, NaCl, atoms of sodium are located at the corners of a cubic array and also at the center of each square face. Bragg reflections from the planes of center atoms would interfere destructively with reflections from the planes of corner atoms. Examining the plane face of a cube of rock salt with their spectrometer, the Braggs found no reflection corresponding to an interplanar distance of the cube edge, but a strong reflection for the distance of half the cube edge. An identical array of chlorine atoms interpenetrates the sodium array, as the black squares surround the red squares on a checkerboard. Along the direction of the cube diagonals, the X-rays encounter alternate planes of sodium and chlorine ions. The recorded spacings and intensities of X-ray reflections confirm the model (see Figure 21).

the Pope and Barlow model, the Braggs completely solved the structure of zinc sulfide. Each zinc is at the center of a tetrahedron formed by four of the sulphur atoms located in the same array as the sodiums (or chlorine) of the Barlow rock-salt model. The structure of diamond is the same except that, instead of zinc and sulphur arrays, both sets of atoms are carbon.

After the interruption of the First World War, when the X-ray tubes were requisitioned for military hospitals, nearly all the inorganic structures were determined. Mathematical methods of great complexity were developed, beginning with the efforts of William Duane (1872–1935). While it is possible to derive the structure constants and the lattice directly from the X-ray data using the relatively simple equations of von Laue or Bragg, the positions of the atoms have to be known in order to predict the intensities of the diffraction. The method has to be one of guessing the structure, at least in part. Structure solutions for other than the simplest crystals would have been hopeless except for crystallographic insights in the tradition of Bravais, Pope, and Barlow.

Out of the first X-ray measurements an art of crystal chemistry grew. V. M. Goldschmidt (1888–1947), a Norwegian geologist, following leads of Landé and Wasastjerna, in 1926 tabulated a set of eighty radii of the ions in crystals. The following year Linus Pauling (1901–   ) formulated a set of empirical rules for structure based on the assumption that ionic structures were packings of large spherical anions (negatively charged), with cations (positively charged) distributed symmetrically in the holes between the spheres. Although the radii were only approximate and mineral structures were not always close-packings, and although there was no real theoretical justification for using similar assumptions for other bondings than ionic, this kind of insight was the basis on which the structures of nearly the entire mineral kingdom were determined between the wars and verified by X-ray analysis.

The earliest carbon structures to be resolved by the new X-ray technique were diamond and graphite. The publication of the Braggs' structure for diamond in 1913 was a dramatic confirmation of the tetrahedral configuration of the carbon atom and of all of stereochemistry, which until then had been simply inference. In Ewald's structure for graphite of 1914, carbon atoms linked together in benzene-like hexagonal rings to form planar sheets. Early work by Bragg in the 1920's confirmed the double-ring structure of the organic chemists for naphthalene and the triple ring for anthracene, assigning dimensions to them which were close to those of the rings in graphite (carbon to carbon distance of 1.42×10^{-8}cm.). It was undoubtedly the complexities of organic structures which fostered the development of sophisticated mathematical techniques such as Fourier analysis. In general, if a reasonable approximation to the structure could be assumed (and the old Kekulé structural formulae were just such guides as were required), the new techniques provided methods for progressively correcting the assumptions until an accurate structure, *the* complete, scaled, geometric configuration of the matter within the crystal, was achieved. It was beginning to be possible to think of solving for structures as complex as the proteins.

In the 1920's T. Svedberg (1884–   ) developed the ultracentrifuge as a means of determining the weights of protein molecules. The settling rates of particles suspended in a liquid in a centrifuge depend on their size and weight. Svedberg reported enormous gram molecular weights (17,300 to 9,600,000, in blocks of 34,600 for the smaller protein molecules and 400,-000 for the larger). The molecular weights of amino acid molecules are of the order of 120, so that the smallest of proteins would have something of the order of 100* amino acid residues, while the largest would contain 10,000. In 1947, E. Katschalski (1916–   ) in Israel and R. Woodward (1917–   ) at Harvard first succeeded in synthesizing such polypeptide chains of hundreds of amino acids, but they were all of the same kind.

X-ray diffraction in the period before the Second World War

* The smallest proteins are now known to be of the order of 10 amino acids with gram molecular weights of the order of 1000.

suggested that the fibrous proteins such as collagen, the protein of tendons and connective tissue; keratin, the protein of skin and hair; and fibroin, the protein of silk, were all made of parallel polypeptide chains, either along the length of the fiber or in spirals with axes parallel to the fiber. W. T. Astbury (1898–1961) found two distinct kinds of X-ray patterns corresponding to the normal (α) and stretched (β) forms of keratin and derived a molecular spring model to account not only for the elasticity of hair but, as he pointed out, for movement in general. "In fact the α-β-type of intramolecular transformation accompanying the stretching of myosin and of muscle protein itself *in situ* has been successfully carried out."

Later A. Szent-Györgyi (1893–) showed that the application of ATP solution to the stretched or β-actomyocin fibers of animal muscle causes rapid contraction. ATP is the substance in which the energy of the oxidation of food is stored, so that this discovery explains the mechanism of animal power.

In the period after the Second World War, Pauling and R. Corey (1897–) and their students made the Gates and Crellin Laboratories at Pasadena a world center of structural protein research, finding, among many others, the structures of the amino acids glycine, alanine, serine, threonine, and providing evidence for their *helix* or spiral-configuration model of the protein molecule. Long helical protein molecules with 3.6 amino acids per turn twist about one another like strands of rope to form fibrous proteins, or fold and tangle into the globular molecules of hemoglobin or ovalbumin.

The first determination of the arrangement of the different amino acids in a chain was accomplished by Frederick Sanger (1918–), who in 1945 devised a method of removing the end amino acids of the chain, one at a time. Each chain had to begin at one end with a COOH group and end with an NH_2 group. Sanger found a reagent that would combine with the amine group and retain this bond while the rest of the chain was breaking down. The amino acids were identified by paper chromatography, a highly simplified new analytic technique. In solutions with reagents the amino acids give characteristic color reactions. When the solutions are allowed to spread through blotting paper, the different amino acids spread at different rates and thus become separated. The procedures and locations of the spreading stains are standardized with known amino acids of various concentrations.

Sanger and Hans Tuppy broke the insulin molecule into two peptide chains, one with 21 and the other with 30 amino acids. By digesting the chains into smaller and smaller fragments and separating the end amino acids, one at a time, they succeeded—some eight years after Sanger's original work—in completely determining the order of the 51 amino acids of the insulin molecule. In the same year, 1953, V. du Vigneaud (1901–) completed the synthesis of oxytocin, a hormone which causes contraction of the muscles of the uterus. It is made of nine amino acids, eight of them different, and in a definite sequence. The order of the 121 amino acids of *ribonuclease* was found in 1959. The order of the 39 amino acids of the hormone ACTH is now fully determined. The order of the 574 amino acids of hemoglobin, with a gram molecular weight of 64,500, has been determined. What is more, after nearly twenty-two years of painstaking research, M. F. Perutz (1914–) and his colleagues have located each of the 574 amino acids in four chains embracing oxygen-carrying (heme) groups. "In 1937," Perutz writes,

> a year after I entered the University of Cambridge as a graduate student, I chose the X-ray analysis of hemoglobin, the oxygen-bearing protein of the blood, as the subject of my research. Fortunately the examiners of my doctoral thesis did not insist on a determination of the structure, otherwise I should have had to remain a graduate student for twenty-three years.

The structure of the individual chains was almost the same as the structure of the single chain of *myoglobin,* which had been worked out in 1957 by John C. Kendrew and others. In homoglobin, two β-chains of helical segments with irregular interruptions, and two α-chains with a slight variation in their convolution, are placed as if at the alternate corners of an almost perfect cube. The oxygen-bearing heme groups are in four separate pockets, one in each chain, symmetrically disposed

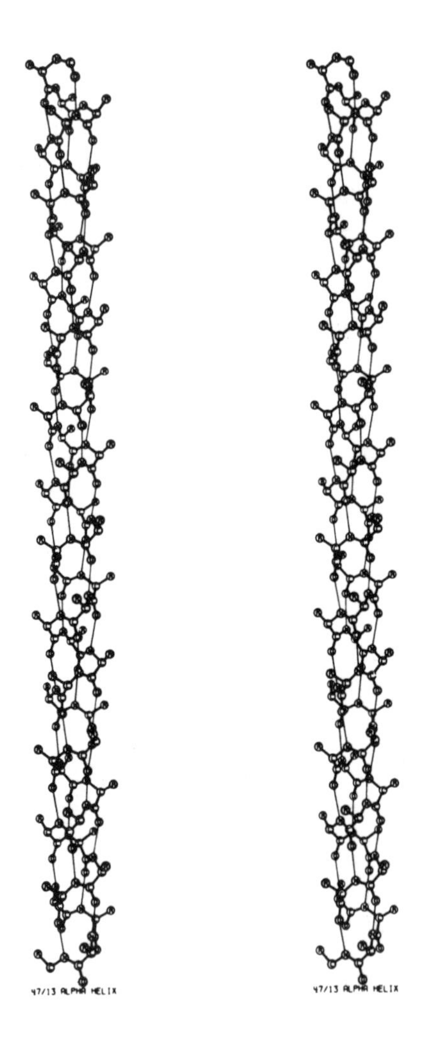

Figure 22. A Polypeptide Chain
The spiral linkage of amino acids to constitute a part of a protein may be viewed in three dimensions with a stereo viewer or by placing a card about eight inches wide between the two images. The drawing is by computer programmed by C. K. Johnson, Oak Ridge National Laboratories.

about the center. The completed molecule is almost a sphere.

If these protein syntheses and structure determinations seem efforts more appropriate to gods or demons than men, even these are overshadowed by the successes of the biochemical attack on the mysteries of the genetic mechanism. "All organic substances occurring in nature may be classified in three large groups," wrote P.A.T. Levene in 1931:

> To the first belong substances found universally in all biological structures, cellular or otherwise; such are proteins, carbohydrates, and fats. The second group contains substances characteristic primarily of the cell; among these stand in the first place nucleoproteins. . . . Finally, the third group embraces substances elaborated by the living cell and discharged by it into the surrounding medium . . .

The principal component of the nuclei of cells is nucleoproteins, the nucleic acids, first separated by Miescher in 1869 and then by Hoppe-Seyler at Tübingen. A quarter-century ago Levene resolved nucleic acid into *nucleotide* blocks combining unit nitrogenous, carbohydrate, and phosphoric parts. The nucleotides play the role of the amino acids in proteins, linking together to form the enormous chain molecules of the nucleic acids.

F. Miescher (1844–1895), of Basel, reported in 1869 to Felix Hoppe-Seyler (1825–1895), the founder of modern biochemistry, that he had succeeded in extracting an indigestible *nuclein* from the nuclei of pus cells by a combination of acid digestion and mechanical separation. More strongly acidic than proteins, it was unusual in containing a quantity of phosphorus. Hoppe-Seyler and his students repeated and confirmed Miescher's work and went on to isolate a second nuclein from yeast cells and similar residues from casein and egg yolk. Miescher obtained his thymonucleic acid (as his nuclein was later called) in exceptionally pure form from the sperm of salmon. From the sperm nuclei he separated a salt of nucleic acid with a nitrogenous base which he named *protamine*. The nucleic acid itself, when subjected to acid hydrolysis—first performed successfully by Albrecht Kossel (1853–1927) in 1891—

yielded four distinct residues which were identified as thymine (isolated by Miescher in 1893), guanine (isolated by Piccard of Basel at Miescher's request in 1874), adenine (discovered by Kossel in 1885), and cytosine (Kossel and Neumann in 1894).

Thymine and cytosine were identified as hexagonal ring structures with two nitrogens within the ring, the so-called *pyrimidine* ring.* Adenine and guanine structures were built on double rings, one ring hexagonal, the other pentagonal. This characteristic double ring is called the *purine* ring.* The same two purines, cytosine and a thymine-like pyrimidine, uracil, were found in equal proportions in the yeast nucleic acid of Hoppe-Seyler by Levene in 1909. The equal molecular proportions of the four nitrogen compounds in thymonucleic acid had been demonstrated by Steudel (b. 1871) three years before.† Uracil had been first separated from yeast or plant nucleic acid by Ascoli in 1900.

Phoebus Aaron Theodore Levene (1869–1940) studied at the Imperial Military Medical Academy of St. Petersburg under Borodin (who was a doctor as well as a composer) and also

---

*

```
       OH                    NH2                NH2
       |                     |                  |
       C     CH3             C                  C       N
     //  \  /              //  \              //  \   /   \\
    N     C               N     CH           N     C       CH
    |     ||              |     ||           |     ||      /
    C     CH              C     CH           CH    C     /
HO/  \\  /            HO/  \\  /              \\  /  \ NH
       N                     N                  N

thymine (pyrimidine)   cytosine (pyrimidine)   adenine (purine)

            OH                         OH
            |                          |
            C      N                   C
          //  \   /  \\              //  \
         N     C      CH            N     CH
         |     ||     /             |     ||
         C     C     /              C     CH
    NH2/  \\  /  \ NH           HO/  \\  /
            N                          N

       guanine(purine)          uracil(pyrimidine)
```

† The molecular proportions are *not* equal for all four nucleotides, but only for adenine-thymine and guanine-cytosine (Erwin Chargoff, 1948).

under Pavlov. To escape the rising anti-Semitism in Russia he immigrated to New York's Lower East Side in 1891 and began to practice medicine. Representative of a generation of eastern European immigrants, he was driven by a restless curiosity from the practice of medicine to the mastery of chemistry and the problem of the nucleic acids.

At this time Simon Flexner directed the Rockefeller Institute, originally organized to provide a center of scientific (medical) research in America that would rival the great laboratories of central and eastern Europe—Kossel's at Marburg, Fischer's at Berlin, Hoppe-Seyler's at Tübingen. In 1905, impressed by Levene's early papers on nucleic acids, Flexner appointed him an assistant in the Institute, which from then on was to become a world center of nucleic acid research.

Levene and his coworkers identified the carbohydrate components of the nucleic acids as *pentoses,* or five-carbon sugars. They were able to show that the pentose of yeast or plant nucleic acid was *d*-ribose—right-handed—while the pentose of the thymus or zoonucleic acid was similar but without one oxygen—*deoxyribose.* The two nucleic acids are called RNA (ribonucleic acid) and DNA (deoxyribonucleic acid) for the sugar components.

The sugar ring and a nitrogen base combine to form a *nucleoside,* first isolated from nucleic acid by Levene and Jacobs in 1909. Later, Levene wrote:

> The success of the work was due in part to the fact that at that time inosine was quite accessible, inasmuch as the Liebig's beef extract then on the market furnished good yields of the nucleoside. The more recent product contains barely traces of inosine or inosinic acid. For the progress of the chemistry of nucleic acids, it is very fortunate that a true Liebig's extract still existed twenty years ago.

On Delancey Street at the turn of the century Liebig's extract was as easy to obtain as tranquilizers are today in Westchester County. It was the specific used by anxious mothers against the threat of malnutrition. The structure of Liebig's inosinic acid was first investigated by Haiser, beginning in 1895. He was able to show that the molecule was made of a phosphoric acid component united through a pentose sugar to a purine (hypoxanthine). This *nucleotide,* as it was called, was essentially an ester of phosphoric acid with a nucleoside. The nucleoside was inosine, named by Haiser. The nucleic acids were all considered as phosphoric acid esters of nucleosides. In the theory of nucleic acid structure developed by Levene and Jacobs the nucleotides were chain-linked by the sugar of each to the phosphoric acid of the next, with the purine and pyrimidine bases extending out from the chain as tails (see Figure 23). The sugar and acid play the same role in the linkage of the nucleotides as the amine and carboxyl acid groups in the linkage of the amino acids.

By the 1950's, beginning with R. Feulgen (1884–1955), who developed techniques for selective staining, biologists had succeeded in locating the DNA within the chromosomes. T. Caspersson (1910–) had narrowed this down to bands (genes?) within the chromosomes, and RNA had similarly been related to the microanatomy of the cell. Rockefeller Institute teams had used DNA to effect a genetic transformation in bacteria, and other researchers had found that the filterable viruses were nucleoproteins, combinations of protein with DNA or RNA. A partial unraveling of the sequence of nucleotides in the molecule was accomplished. Actually the DNA molecules in a single cell nucleus may represent a sequence of a hundred million nucleotides.

In 1953 F.H.C. Crick (1916–), a crystallographer, and J. D. Watson (1928–), a biologist working at Cambridge, proposed a model of DNA and the genetic mechanism. From X-ray analyses they could show that the basic polynucleotide structure was a left-handed helical sugar-phosphate chain with projecting purine and pyrimidine components—like the Pauling and Corey model of the protein helix. Two of these helices were linked together, the double-ring purines of one bonded to the single-ring pyrimidines of the other (adenine opposite thymine and guanine opposite cytosine). The double helix splits apart when the cell divides, and each separate chain then synthesizes the missing half by attaching opposite nucleotides—a thymine

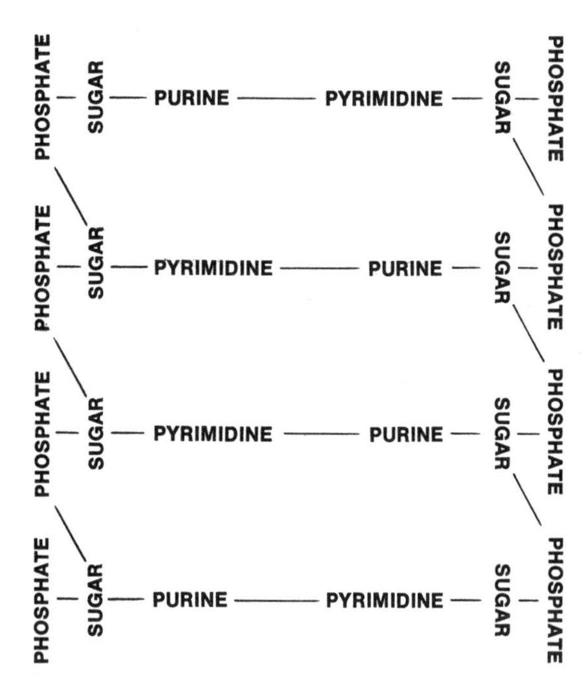

Figure 23. The Nucleic Acid Chain
The Crick and Watson double helix. A pentose sugar-base block forms a *nucleoside*. A phosphate-nucleoside block forms a *nucleotide*. Successive nucleotides are linked by the phosphate of one joining the sugar of the next. Two helical polynucleotide chains link together to form a spiral-staircase-like helix. The purine-pyrimidine bases link together to make the steps, adenine always linked to thymine, and guanine to cytosine, so that the coding sequence of one determines the sequence of the other.

attaching to each adenine, and so on. If all the molecules of a chromosome split and replicate themselves, the chromosome will be duplicated and with exactly the proper sequence.

In a similar fashion, RNA is supposed to act as the catalyst for the formation of particular proteins. Each half of the DNA double helix acts by itself as a template for replication, a mold which completes itself by selective bonding with nucleotides. M. B. Hoagland (1921–) proposed that the DNA carries the genetic information which catalyzes the assembly of RNA from nucleotides. The RNA in turn catalyzes the formation of proteins from amino acids by selectively linking to these until a properly sequential polypeptide is complete, when it is released and a new one begun. Severo Ochoa (1905–) and Arthur Kornberg (1918–) used the opposite procedure, employing the catalytic action of enzymes (proteins) to bring about the synthesis of RNA and DNA from nucleotides. Hoagland had already shown experimentally the selective affinity of nucleic acid fragments for amino acids.

In 1967 Kornberg (with M. Goulian and R. L. Sinsheimer) used single-stranded DNA from a natural virus as a template for the synthesis of mirror strands of DNA from off-the-shelf materials. The synthetic DNA infected living cells, giving rise to virus particles that were essentially reflection images of the original virus and with the full infectious power of the original virus. In addition to this, the synthetic mirror strands were themselves used as templates for the direct synthesis from off-the-shelf materials of fully infectious, single-stranded DNA of the original virus.

One basic problem in this mechanistic picture of heredity is that of information. Even the hundred million or so atoms in the sequence of nucleotides in the cell could hardly be expected to carry all of the innumerable bits of information that go to make up the complexity of a man. A simpler problem is the relationship between the nucleotides and the twenty amino acids. George Gamow (1904–), the nuclear physicist who should be described more properly as polymath because of the diversity of the fields he has enriched, suggested that if the nucleotides acted in groups of three, they could encode exactly twenty different keys, one for each of the amino acids. There are four ways in which all three in a group may be different, one for each of the four nucleotides which has to be omitted if only three are to be selected. There are four times three or twelve ways to make up a group of three from only two of the four nucleotides, and there are four ways to make up a group of three of all the same kind of nucleotide. However, the association of each of the amino acids with a definite triple-nucleotide group is still but a suggestion, the validity of which remains to be established.

We are caught in a paradox. On the one hand, our theoretical physics has rid itself of all traces of mechanism with the relativistic theory of gravitation. We have substituted the uncertainty principle for determinism, or at least for the rigid Laplacian belief that the whole of the past and the future could be derived inexorably from the particles, their configuration and their motions. Even the conservation laws have not survived the discovery of radioactivity, so that the ancient alchemical dream of transmutation is as ordinary as sunlight.

Yet the phenomena of reproduction and sexuality emerge, as we penetrate to deeper levels of understanding, as the purest mechanism of all. In the coding of the nucleic acids we feel again the triumph which was Newton's when finally the mechanism of the Macrocosm stood clear. If the seventeenth century penetrated to the mechanism of the heavens, so we stand before a curtain rising on the mechanism of man, the Microcosm. In this most complex of patterns, the double helices of the DNA and RNA, life itself seems an expression of the form and number of the Pythagoreans. The old Cartesian dualism is ended, and between the organic and the inorganic kingdoms we find continuous passage.

15

The Structure of the Atom

WHERE HAS MY FOOLISHNESS LED IN THE EFFORT TO PRESENT A MERE NOTHING, FOR I HAVE CREATED ALMOST A WORLD OUT OF THIS NOTHING AND LEAVING THE SOUL OF THE SMALLEST BEING, I HAVE ASCRIBED THE SOUL OF THE LARGEST, THE EARTH, TO A SNOWSTAR?

—KEPLER, 1611

Once the integrity of the atom had been breached, a new approach to intra-atomic structure was possible. In the development of atomic crystallography the "texture," as Boyle called it, of matter was explained in terms of the number, configuration, and motions of aniomeric particles. Not simply the sheets of atoms of carbon sliding one past another to make the lubricant graphite, but the very processes of generation and heredity were being explained in mechanical terms.

Could the same mechanism be traced within the particles themselves? If the *minima* are now observed to break down in radioactive processes, what of the still smaller particles into which the minima are broken? Are they in turn duplicates on a smaller scale, replicating in their mechanism the larger mechanism of the minima? Are they the true "atoms," the final hard round massy particles, like Euclidean points, without parts, permanent, indivisible? Or are they like the picture of the cereal box within which we discern a minuscule picture of a cereal box within which we discern still another picture of a cereal box and so on ad infinitum?

> . . . The approximate value for the central charge of the atom of gold . . . is about that to be expected if the atom of gold consisted of 49 atoms of helium each carrying a charge 2e. This may be only a coincidence but it is certainly suggestive in view of the expulsion of helium atoms carrying two unit charges from radioactive matter . . .

Rutherford wrote from Manchester in 1911.

Two years before, he had assigned the task of measuring the scattering of α-particles by metal foils to Hans Geiger (1882–1945), who was working in his laboratory, and to Geiger's student E. Marsden (1889–). From estimates of the Avogadro number and the thickness of a given weight of foil (about 4×10^{-5} cm.) it could be determined that an atom of gold was of the order of an Angstrom unit (10^{-8} cm.) in radius. If the charge of the atom were spread throughout the whole volume of the atom, the electric field would not be strong enough to turn α-particles expelled from radon, which were estimated to travel at about 1/10th the velocity of light. In other words, the α-particles moved with a speed of ten to twenty thousand kilometers per second. Geiger and Marsden were astonished to find that some of the α-particles were reflected back from the thin gold foil.

"A simple calculation based on the theory of probability shows that the chance of an α-particle being deflected through $90°$ is vanishingly small," Rutherford wrote.

J. J. Thomson had pictured the atom as a number of negatively charged particles moving in circular orbits in a sphere of "uniform positive electrification," like the seeds in a pumpkin. In the same year, 1904, Nagaoka had proposed a model for the atom like the planet Saturn—a central mass surrounded by rings of rotating electrons. Rutherford, starting with this, showed that he could account for the pattern of scattering of the α-particles if all the positive charge of the gold atom, which he calculated at about 100e (actually 79e), was concentrated in a nucleus at the center of the atom. The nucleus was to be something less than 10^{-12} cm. in radius, so that the volume of the nucleus with all the positive charge of the atom would be of the order of 10^{12} times smaller than the atom. By analogy, if we fired a machine-gun burst of steel bullets at a given weight of steel pressed into a thin stamping, we would expect the bullets to pass through easily. However, if the same weight of steel were concentrated in a single small lump, then we would expect a few of the bullets encountering the lump to be deflected at sharp angles. With the further assumption that the mass of the atom was also concentrated in the nucleus, this became the Rutherford planetary model of the atom. Unlike other speculations—even those like Nagaoka's, which were mathematically based—this planetary model attracted great interest, because the scattering experiments seemed to be a direct measure of the intra-atomic configuration.

Rutherford's model was nevertheless unsatisfactory. Inference from radioactive phenomena suggested that there must be separate positive and negative charges. Coulomb's Law required that these be in motions and configurations which would keep them separate. "The forces acting on the electrons in the atom-model of Thomson," Niels Bohr wrote in 1913, "allow of certain configurations and motions of the electrons for which the system is in a stable equilibrium; such configurations, however, apparently do not exist for the second [Rutherford] atom-model."

Yet it was the Rutherford atom model which emerged from the only experiments aimed directly at the geometry of the atom. Niels Bohr (1885–1962) came to England from Denmark in 1911 and joined Rutherford's group. As Bohr analyzed the problem, the single planetary electron of the hydrogen atom would have to travel about the positively charged nucleus with a great enough velocity for the centrifugal force to balance the Coulomb force of attraction. But according to the theory of electromagnetic radiation, as Maxwell had proposed it and Hertz had experimentally confirmed it, ". . . the electron will approach the nucleus describing orbits of smaller and smaller dimensions . . . at the same time as the whole system loses energy." This would be in the form of electromagnetic radiation (light) of continuously changing frequencies. If Rutherford's model was correct, matter would spontaneously incandesce and collapse. If Thomson's model was correct, there would be no high-angle scattering of α-rays. Bohr found a solution to the contradiction in the new quantum theory of radiation.

> . . . The essential point in Planck's theory of radiation is that the energy radiation from an atomic system does not take place in the continuous way assumed in the ordinary electrodynamics, but that it, on the contrary, takes place in distinctly separated emissions, the amount of energy radiated out from an atomic vibrator of frequency ν in a single emission being equal to $\tau h\nu$ where τ is an entire number. . . .

He proposed that the system with the electron in orbit of radius r was in a stable or stationary state. The potential energy of this state was equal to the energy of bringing a negative charge from an infinite distance to the distance r from the positive nucleus, and the kinetic energy was calculated from the velocity in orbit required to counterbalance the Coulomb force of attraction between opposite charges. Energy could be radiated out or absorbed only in discrete units or quanta of value $h\nu$. Considering the infalling electron as changing from a state of rest to a state of rotation about the nucleus with a definite frequency (ω) and energy, Bohr assumed that the infalling electron would emit "a homogeneous radiation . . . of a frequency ν, equal to half the frequency of revolution of the electron in its final orbit. . . ." The quantum of emitted radiation Bohr set as equal to $\tau h\nu$, or $\tau h\omega/2$. Knowing the mass and charge of the electron and the value of Planck's constant h, it was possible to solve for r, which came out to be 1, 4, 9, 16, etc., x .5 x 10^{-8} cm., as τ was 1, 2, 3, 4. etc. With r known it was possible to calculate the frequencies of the radiation quanta emitted or absorbed as the electron changed from one stable state or orbit, τ_1, to another, τ_2. Bohr showed that the frequency of the quantum must be equal to

$$\frac{2\pi^2 me^4}{h^3}\left(\frac{1}{\tau_2{}^2}-\frac{1}{\tau_1{}^2}\right). \qquad (1)$$

> We see that this expression accounts for the law connecting the lines in the spectrum of hydrogen. If we put $\tau_2 = 2$ and let τ_1 vary, we get the ordinary Balmer series. If we put $\tau_2 = 3$, we get the

series in the ultra-red observed by Paschen and previously suspected by Ritz. . . .

Convinced by this success of the physical reality of his model, Bohr recast his ideas as deduction from a fundamental assumption; the angular momentum of the electrons in atomic systems must be quantized (which is to say equal to a whole number times $h/2\pi$):

$$Mvr = nh/2\pi. \qquad (2)$$

In the second part of this extraordinary article, Bohr, who was only twenty-eight at the time, outlined a scheme for the structures of all the elements of the periodic chart. From the

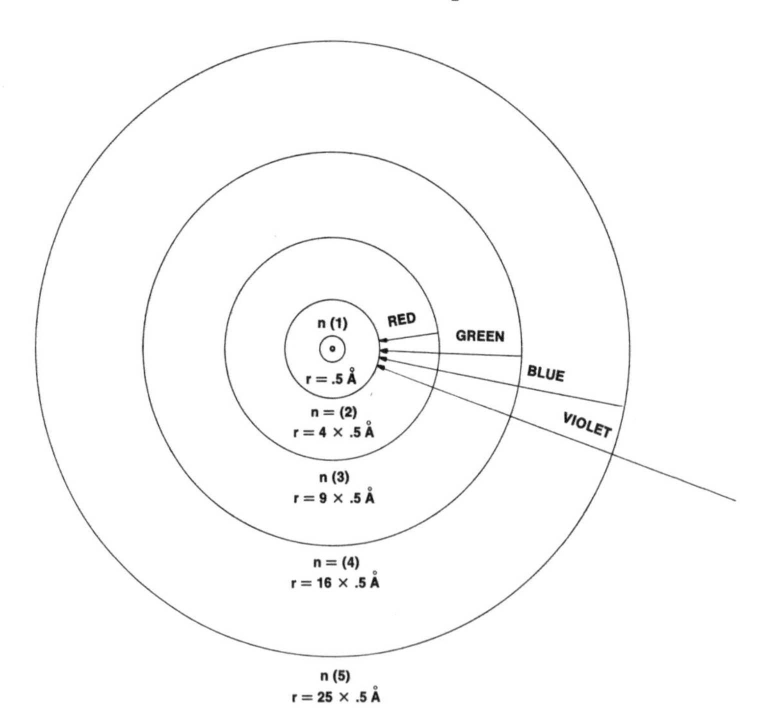

Figure 24. The Bohr Model of the Hydrogen Atom
Electron exchanges to the second Bohr orbit are accompanied by the emission of photons of visible light (the red, green, blue, and violet of the Balmer Series).

experiments of Rutherford, Geiger, and Marsden on the scattering of α-particles and the experiments of C. G. Barkla (1877–1944) on X-ray scattering, he came to the hypothesis of A.v.d. Broek that the neutral atom contained the same number of electrons (Z) as "the number which indicates the position of the corresponding element" in the periodic chart. He then used his quantum assumption (2) to obtain the distributions of the Z electrons in the rings of the Zth element. Of the various possible configurations which would satisfy the conditions of the two hypotheses, he used the chemical properties to make the best choice. The configurations as Bohr drew them (but in modern notation) are:

Atomic Number	Element	K 1s	L 2s 2p	M 3s 3p 3d
1	H	1		
2	He	2		
3	Li	2	1	
4	Be	2	2	
5	B	2	3	
6	C	2	4	
7	N	4	3	
8	O	4	2	2
9	F	4	4	1
10	Ne	8	2	
11	Na	8	2	1

etc.

The details of Bohr's model were corrected in the next decade by his own efforts and by the ideas of Sommerfeld, Pauli, Ehrenfest, Uhlenbeck, Goudsmit, and others. Bohr and Sommerfeld established a second quantum integer (l) representing the mechanics of quantized ellipses, to supplement the circular orbits of the primitive theory. The integral values of l go from 0 to n-1, where n is the principal quantum number. They are represented by *s, p, d,* and *f,* etc., standing for the characteristics of the detailed line spectra, *s*harp, *p*rincipal, *d*iffuse, *f*undamental, etc. A third quantum number to account for magnetic effects, m_l, takes on values from $-l$ to $+l$. W.

Pauli (1900–1958) observed a "two valuedness not described classically" for the allowed energies of atomic systems. G. Uhlenbeck (1900–) and S. Goudsmit (1902–), students of Ehrenfest at Leiden, attributed to the electron a spin with an associated magnetic moment either parallel or antiparallel to the angular momentum. This was the fourth or spin quantum number. Pauli's two-valuedness became the exclusion principle, according to which no two particles in the same system may take on identical values for all four quantum numbers.

The application of the exclusion principle to the quantum numbers yields the periodic table up to Z(19) before the first discrepancy appears. For Z (1), n is 1; l takes the value 0 or s, m_l is therefore 0, and m_s (the spin number) is either $+\frac{1}{2}$ or

n	l	m_l	m_s
1	0	0	$\pm\frac{1}{2}$
2	0	0	$\pm\frac{1}{2}$
	1	-1	$\pm\frac{1}{2}$
		0	$\pm\frac{1}{2}$
		$+1$	$\pm\frac{1}{2}$
3	0	.	.
	1	.	.
	2	.	.
		.	.
			.

$-\frac{1}{2}$. For Z(2), helium, each of the two electrons takes one of the two spin numbers.

For the model in its later stages, the first two of the Z electrons of the Zth element form an inner energy state, the 1s state or helium core. The next two form the 2s state through Z(4), which is beryllium. The next six electrons of the elements boron through neon enter the 2p state characterized by three possible values of m_l and spin numbers plus or minus $\frac{1}{2}$. With neon, Z(10), written $1s^2 2s^2 2p^6$ (with superscripts standing for the number of electrons), a chemically stable core is again reached. Bohr used the ionization potentials—that is, the calculated energies required to add or remove electrons from a neutral atom—as the criterion by which to distinguish the true configuration from the various possibilities consonant with his quantum restrictions. Spectroscopy—the presence or absence of the appropriate frequencies from the spectra, and the energies of their excitation—provides the final criteria.

The letter designations K, L, and M for the states with principal quantum number $n = 1, 2$, and 3, were originally assigned by Barkla to the characteristic lines of the X-ray spectra. H.G.J. Moseley (1887–1915), following the discovery by W. H. Bragg of the characteristic X-ray spectra, demonstrated that their frequencies were a function of their atomic number Z, and independent of atomic weights, etc. This discovery of Moseley's was the justification of the periodic chart. All of Mendeléev's work, the numerical laws of classical chemistry, the flood of empiricism which characterized the tough-minded scientist of the nineteenth century, was finally to be explained in these developments of modern physics. Moseley was one of Rutherford's group, with Geiger and Bohr. He was killed at the age of twenty-eight, in the landings at Suvla Bay, Gallipoli, in 1915. His colleague Geiger spent four years as a soldier in the same war—on the other side.

In the further development of physics, specifically in wave mechanics and the equivalent quantum mechanics, the literal realism of the Bohr model is completely superseded. The physicist no longer speaks of the electron in the second orbit, for example, but rather of the electron density characteristic of the $2s$ or $2p$ state. The K, L, and M shells of G. N. Lewis (1875–1946) and I. Langmuir (1881–1957) are no longer considered guides to the spatial configurations of the electrons about the atoms. They are useful and used, as valences are useful and used, but in the same sense as one speaks of the sun rising and setting, long after we have accepted the Copernican view that it is the earth itself which moves.

Lewis and Langmuir undertook to account for the chemical bond, proposing that in one bond type (the homopolar or covalent bond), such as the chlorine-chlorine bond in the diatomic molecule, a pair of electrons are shared by two nuclei. Each of the chlorine atoms has seven electrons in its M shell, one less than the eight of the inert gas argon. If each contributed its seventh electron to a shared pair, both would achieve the argon configuration. In ionic compounds, as had been suggested by W. Kossel (1888–1956), an atom with configuration close to the stable configuration of the completed outer shell of an inert gas would gain or lose an electron, becoming a negative or positive ion which would then bond together by ordinary Coulomb forces. M. Born (1882–) and A. Landé (1888–) showed that if the Coulomb energies between all the ions in a crystal were summed according to a method originally proposed by E. Madelung (1881–), and if the force of attraction between oppositely charged ions was assumed to be reversed as the ions approached very closely to each other, the energy of the crystal could be calculated with results in excellent agreement with the energies required to dissociate the crystal.

The properties of conductivity, opacity, and ductility of the metals made them a special case. Drude (1863–1906) and Lorentz (1853–1928) proposed that the electrons within the metals were free, forming a kind of gas. The motion of free electrons in an applied field agreed quantitatively with the calculated resistivities. Simply treating the free electrons as classical kinetic-molecular particles, the theory accounted for the high conductivity as the motion of the current of electrons toward the positive side of an applied field. Particularly, the theory predicted that the ratio of the electrical to the thermal conductivities should be the same for all metals (Wiedemann-Franz Law). But such a gas of particles should have added three calories to the molar heat capacities, and nothing like this was observed.

In 1928 Sommerfeld applied the restrictions of the quantum theory to the free electrons, describing them in terms of their wave properties. Because Pauli's exclusion principle provided that no more than two electrons in the whole system could occupy the same energy state at the same time—a restriction much like the materialist assumption that no two bodies may occupy the same space at one time—only a minute proportion of the electrons were free to move to higher energy states at ordinary temperatures; therefore their contribution to the specific heat was negligible. In 1928 also, Bloch changed the model to one of electron waves in a periodic medium provided by the positive cores of the atoms in regular crystalline array.* Just as X-rays either penetrate or are reflected by the periodic distribution of the atoms within the crystal, the electron waves exhibit the same mathematical and geometric relationships.

A remarkable development of the theory of the solid state grew out of these researches. The physical and chemical properties of solids were to be explained in terms of the propagation of waves in periodic structures. "Waves always behave in a similar way, whether they are longitudinal or transverse, elastic or electric," wrote Léon Brillouin (1889–). Brillouin showed that the energies of waves in solids were restricted to sharply defined bands, bounded in zones in three dimensions by polyhedra related to the crystal structure, and geometrically similar to the natural growth form of the actual crystal.

Pauling, trying to account for the stability of structures such as the benzene ring (see page 138), with alternately double- and single-bonded carbons, proposed a new concept, that of the *resonant bond,* following a suggestion of Heisenberg. From the quantum-mechanical point of view, a structure such as Kekulé's benzene ring, consisting of atoms linked together by shared electrons, is represented by a wave function. If the arrangement is in any way changed, the structure must be represented by a slightly different wave. In the same way that two waves of light interfere with each other, either increasing or decreasing the amplitude of the resultant wave according to the degree to which their crests and troughs reinforce or cancel each other, the waves of the two arrangements are added together to give a third wave function. If the energy for the system calculated with the third wave is lower than that calculated for either of the waves for the two structures, the system is said to *resonate* between the two structures, or to be a *resonance hybrid* of the two.

The resonance hybrid is stable relative to the first Kekulé structure, because to convert it from the hybrid state to the first

---

* Bloch's model is directly analogous to the eighteenth-century postulate of Boscovich (see Figure 12).

structure would require energy or work equal to the difference in the energies calculated with the respective wave functions. Similarly, it is stabilized with respect to the second structure by an amount equal to the same energy difference. This extra stability of the system, relative to the most stable of the configurations chosen, Pauling calls the resonance energy. He describes the hybrid as resonating between the two or more simpler structures from which the wave function is constructed.

> . . . There is no single valence-bond structure that accounts satisfactorily for the properties of benzene. The simple description of benzene that is given by the theory of resonance involves two valence-bond structures, the two Kekulé structures  and  . These two structures must be superimposed, fused together, to represent the molecule of benzene, with consideration given also to the stabilizing effect of resonance—that is, the benzene molecule does not have a structure midway between two Kekulé structures, but rather a structure that is changed from the intermediate structure in the way corresponding to energetic stabilization. . . .

These applications to structural chemistry are actually, as Pauling remarks, empirical extensions of valence-bond theory, rather than deductions from quantum mechanics:

> . . . during the 35 years since the Schrödinger equation was discovered only a few accurate nonempirical quantum-mechanical calculations of the properties of substances in which the chemist is interested have been made.

Intellectually, the theory of resonance in chemistry must be considered as the extension to chemical-structure theory of the quantum-mechanical theory of the covalent bond initiated by Condon (1902–), Heitler (1904–), and London (1907–). As originally designed to account for the diatomic hydrogen molecule, wave functions are written for the structure in which each electron is associated with its proton nucleus, and for the alternative structure in which the proton nuclei exchange electrons. Calculations then show that about 80 percent of the observed energy of the electron-pair bond is the resonance energy of the hybrid of the two structures. If ionic structures—a negative hydrogen ion with both electrons attached to the same nucleus and a positive hydrogen ion, and the mirror structure with the ions reversed—are considered, this ionic resonance energy accounts for another 5 percent of the observed energy, the remaining 15 being accounted for by other, more complicated interaction. This is the sense in which a chemical bond may be described as partially covalent and partially ionic, or partially metallic. Weak chemical bonds between electrically neutral particles are called van der Waals bonds. These are the bonds which hold together organic crystals or join fully satisfied layers such as the silicate sheets of talc or the layers in graphite. These bonds also may be explained by the composition of wave functions.

The value of these concepts does not lie in the calculations, which proceed only with great difficulty and then yield energy values which in any event are already known through experiment, but in their enormous fruitfulness in supplying models, even mechanical models (although whether the quantum mechanics should be considered as mechanical is questionable). With these models a much more sophisticated structure theory of organic chemistry is possible. Because of this theory the determination of still unknown structures is facilitated; from the relationships between the theory and the properties, synthesis and what may be described as chemical architecture are advanced immeasurably.

16

CORPORA AD LECTOREM
(Salute to the Reader)

THAT SWEET ELUSIVE FRUIT, SO DELECTABLE,
PHILOSOPHERS WERE ONCE COMPELLED TO SEEK,
IT WAS IN OUR BEHALF, TO FEED OUR INTELLECT.

—LEONARDO, THE CODEX M

After he had persuaded Leonardo, that "ineffable left hand," to illustrate the manuscript of *De Divina Proportione,* Fra Luca Pacioli placed the artist's *terzina* on the final page, a summary of what had gone before.* It was 1498. In another year Milan was overrun by the French. The artist and the mathematician left together for Florence. The brief period of flowering of the court of Milan was over. If their labors had been arduous, they had been well rewarded by the taste of that sweet elusive fruit. They had constructed a model of the world out of number and proportion and the five perfect polyhedra of Plato. After that the world was proportionate, reasonable, and perfect, in the way that nature has of imitating art and science. Later critics such as Duhem might rail against the use of models as crutches for "weak English minds," but a generation which has seen the synthesis of proteins and seriously proposes to unravel the genetic code is too curious to be deterred by the code of chivalry which the purists would impose upon science.

For there are the purists, and to some extent this describes the

* Pacioli, p. 47.

theoretical physicists, whose concern is with the esthetic side of science. These are the men who seek for their work the mathematician's accolade of elegant. In their view, science is a search for order, for full and harmonious solutions to Plato's problem, its task the reconciling of the apparent disorder of things with the innate conviction that the world is invested with a predetermined harmony of number and form.

For Lavoisier or Duhem as for the Newton of the *Principia,* the world was to be comprehended in a single, logical, deductive system. Science is a great contest, the scientist's code as rigid as that of the Samurai. But for the chemist caught in the fascination of a new synthesis, or the crystallographer establishing the reality of the benzene ring, nature is a harlot, her secrets to be wrung from her without compunction. Where reason fails, the computer may yet decide the issue. If there is no other way to find the value of π, we shall cut a circle from paper and weigh it. Unable to deduce the structure of a crystal from the X-ray pattern, we shall guess the structure and see if our postulated structure has the properties of the real crystal.

These are the curious scientists, the opportunists of science for whom philosophy must always be subordinate to experiment, for whom nature is capricious, inconsistent, even vain, but always and in every aspect fascinating. Like Priestley or the Newton of the *Opticks,* they do not shrink from the boldest speculations, the rashest of conclusions. In the history of chemistry perhaps even more than in the history of physics, this interplay of logic and necessity, of cautious philosopher and rash theologian, is evident.

The truth of nature as well as the true nature of matter is no more within our grasp than it was for Thales of Miletus. The chemist, like the Ionian Greeks, inquires, "What are things made of?" The answers of the Ionian Greeks were simple and direct, with all of the charm of the childhood of the race. The vast intellectual construction which is the modern science of matter grew out of the first answers.

We have lost the innocence of the ancients, who walked to the Agora to discourse of philosophy. We have lost their moral innocence with our bombs and our nerve gas and our germ warfare—as well as our plastics and luncheon meat and jet fuel. But we have gained immeasurably in that we have achieved the power to put an end to war and famine and pestilence and poverty. We have gained intellectually because it is not in the nature of man to rest content with the song of a child. The condition of man is this: that he is in a state of tension with the unknown. It is not vanity which motivates true science, nor greed, nor yet ambition, although these being part of the nature of man must play a part even here. It is the state of tension which drives us; we find grace in the simple fitness of the periodic chart or the intertwined helices of DNA. Aware that these too are constructions, the one to be explained by the form and number of the quantum theory, the other in terms not yet conceived, we find in them all the complexity, all the harmony of the thought of a mature and cultivated society. These are the best things of which man is capable. If we can also, like the contemporaries of Thucydides, destroy ourselves in the senseless pursuit of wealth and the drive toward the dominance of man by man, we have shown in this most beautiful and intricate science of matter the heights to which the mind of man can rise. The purpose of this enterprise is indeed, as Hooke once said, to lighten the labors of man's hands and to soften the ancient curse of Eden, but also, because it is the nature of man to find joy in pure thought, to exercise to the fullest the minds which we were given, on the world in which we were born.

Bibliography

*Agricola (Bauer, G.), 1546, *De Natura Fossilium;* 1955, *Textbook of Mineralogy* (Bandy, M.C., and Bandy, J., transl.) (New York: Geological Society of America).

*———, 1556, *De Re Metallica;* 1950 (Hoover, H.C., and Hoover, L.H., transl.) (New York: Dover).

*Aristotle, *Physics,* VI, 9; 1929–35 (Wicksteed, P.H., and Cornford, F.M., transl.) (London: Heinemann).

*Arrhenius, Svante, 1887, "Über die Dissociation der in Wasser gelösten Stoffe" (On the Dissociation of Substances in Aqueous Solution), *Zeits. phy. chem., 1,* p. 631.

Asimov, I., 1960, *The Intelligent Man's Guide to Science,* 2 vols. (New York: Basic Books).

*Avogadro, A., 1811, "Essai d'une Manière de Déterminer les Masses Relatives des Molécules Élémentaires des Corps, et les Proportions Selon Lesquelles Elles Entrent dans Ces Combinaisons," *Jour. de Physique, LXXIII,* pp. 58–76.

*———, *Opere Scelti di,* 1911 (Turin: Accademia Delle Scienze di Torino).

*Bacon, F., 1620, *Novum Organum;* in *Great Books of the Western World,* vol. 30, 1952 (Chicago: Encyclopedia Britannica).

*———, 1627, *The New Atlantis; ibid.*

Bailey, C., 1964, *The Greek Atomists and Epicurus* (New York: Russell & Russell).

*Barlow, W., 1883, "Probable Nature of the Internal Symmetry of Crystals," *Nature,* December, pp. 186, 205, 207.

Basham, A.L., 1954, *The Wonder That Was India* (New York: Grove).

* Primary source.

*Beguyer de Chancourtois, 1862, "Mémoire sur un classement naturel des corps simples ou radicaux appelé 'vis tellurique,'" *Comptes Rendus, 54,* pp. 757–761, 840–843, 967–971.

*Bernouilli, D., 1738, *Hydrodynamica;* in Magie.

Berthelot, M.P.E., 1902, *La Révolution Chimique: Lavoisier* (Paris: Alcan).

*——, *Collection des anciens alchimistes grecs,* 1963 (London: Holland).

*Berzelius, J.J., 1814, "Essay on the Cause of Chemical Proportions, . . . Together with a short and easy method of expressing them," *Ann. of Phil., 3,* pp. 51–52, 362–363; in Leicester and Klickstein.

Biographie Universelle, 1841, 2nd ed. (Weiss, M., ed.) (Paris: Furne).

*Birch, T., 1756–57, *History of the Royal Society of London,* 4 vols. (London: Millar).

*Black, J., 1754, *Experiments upon Magnesia Alba;* 1893 (London: Alembic Club #1).

*——, 1803, *Lectures on the Elements of Chemistry,* Published by John Robinson (Edinburgh: Creech); extract in Magie.

*Bohr, N., 1913, "On the Constitution of Atoms and Molecules," I, II, III, *Phil. Mag., 26;* 151, p. 1; 153, p. 476; 155, p. 857.

*Boltzmann, L., 1896, *Lectures on Gas Theory;* 1964 (Brush, S.G., transl.) (Berkeley: Univ. of Calif.).

——, 1905, "Memorial to Josef Loschmidt," *Populare Schriften;* 1925 (Leipzig: Barth), pp. 228–252.

*Boscovich, R.J., 1763, *A Theory of Natural Philosophy;* 1922 (Child, J.M., transl.) (Chicago: Open Court).

*Boyle, R., 1772 (Birch, T., ed.), *The Works of the Honourable Robert Boyle;* extracts in Guerlac, 1953.

*——, 1661, *The Sceptical Chymist;* 1949 (London: Everyman, J.M. Dent).

*——, 1662, *New Experiments Physico-Mechanical Touching the Spring of the Air and Its Effects;* in Shamos.

*———, 1666, *The Origine of Formes and Qualities* (Oxford: H. Hall).

*Bragg, W.H., and Bragg, W.L., 1913, "The Reflection of X-rays by Crystals," I, II, *Roy. Soc. Proc.*, A, *88*, p. 428; *89*, p. 246.

*———, 1913, "The Structure of the Diamond," *ibid.*, *89*, p. 277.

*Bragg, W.L., 1913, "The Structure of Some Crystals as Indicated by Their Diffraction of X-rays," *Roy. Soc. Proc.*, A, *89*, p. 248.

*Bravais, Auguste, 1850, *Mémoire sur les systèmes formés par des points distribués régulièrement sur un plan ou dans l'espace* (Paris: École Centrale des Travaux Publique).

*———, 1866, *Études Crystallographiques* (Paris: Gauthier-Villars).

*Brillouin, L., 1953, *Wave Propagation in Periodic Structures*, 2nd ed. (New York: Dover).

*Brown, R., 1827, "On the General Existence of Active Molecules in Organic and Inorganic Bodies"; extracts in Magie and in Guerlac, 1953.

Brown, S.C., 1962, *Count Rumford* (Garden City: Anchor).

Butterfield, H., 1960, *The Rise of Modern Science, 1300–1800* (New York: Macmillan).

*Cahours, A., 1857, "Recherches sur les Acides Amidés des Acides Monobasiques," *Comptes Rendus*, *44*, pp. 567–571.

Cajori, Florian, 1915, "The History of Zeno's Arguments on Motion," *American Mathematical Monthly*, *XXII*, p. 1.

*Cannizzaro, S., 1858, "Sketch of a Course of Chemical Philosophy," *Il Nuovo Cimento*, *vii*, p. 321; 1910 (Edinburgh: Alembic Club #18).

*Carnot, S., 1824, *Reflections on the Motive Power of Fire*; 1960 (Thurston, R.H., transl.) (New York: Dover).

Casson, Stanley, 1939, *The Discovery of Man* (New York: Harper).

*Cavendish, H., 1766, "Experiments on Factitious Air," *Phil. Trans.*, *56*, p. 141; in Guerlac, 1953.

*———, 1789, "Composition of Water," *Phil. Trans., 74,* p. 119; 1950 (Edinburgh: Alembic Club #3).

*———, *The Scientific Papers,* 1921 (Maxwell, J.C., ed., Larmor, J., rev.), 2 vols. (Cambridge: Cambridge Univ.).

Chalmers, A., 1812, *Biographical Dictionary,* 32 vols. (London: Nichols).

*Chevreul, M.E., 1832, "Neues eigentumliches stickstoffhaltiges Princip. im Muskelfleisch aufgefundet," *Schweigger Jour., LXV,* pp. 455-456.

Childe, Gordon, 1954, *What Happened in History* (Baltimore: Penguin).

*Clausius, R., 1850, "On the Motive Power of Heat," *Poggendorff's Ann. der Physik, 74,* p. 368; in Carnot.

Collingwood, R.G., 1960, *The Idea of Nature* (New York: Oxford Univ.).

Conant, J.B., 1952, *On Understanding Science* (New Haven: Yale Univ.).

*Condorcet, Marquis de (Caritat, N.), 1794, "Esquisse d'un tableau des progrès de l'esprit humain"; in *Les Grands Écrivains Scientifiques,* 1902 (Laurent, A., ed.) (Paris: Colin).

Cornford, F.M., 1957a, *Plato's Cosmology* (New York: Liberal Arts).

———, 1957b, *Plato and Parmenides* (New York: Liberal Arts).

*Couper, A.S., 1858, "Sur une Nouvelle Théorie Chimique," *Comptes Rendus, 46,* p. 1157; 1933, "On a New Chemical Theory," *Phil. Mag., 16,* 4, p. 104 (Edinburgh: Alembic Club #21).

Creel, H.G., 1961, *The Birth of China* (New York: Ungar).

———, 1963, *Chinese Thought* (New York: New American Library).

*Curie, P., Curie, M., and Bémont, G., 1898, "Sur une Nouvelle Substance Fortement Radio-active, Contenue dans la Pechblende" (radium), *Comptes Rendus, 127,* p. 1215.

*——, and Curie, S. (Marie), 1898, "Sur une Nouvelle Substance Radio-active, Contenue dans la Pechblende" (polonium), *Comptes Rendus, 127,* pp. 175–178.

Cuvier, G., 1819, *Recueil des Éloges Historiques* (Strasbourg: Levrault).

*Dalton, J., 1805, *Experimental inquiry into the proportion of the several gases or elastic fluids constituting the atmosphere* (Manchester Memoirs, Second series, I); in *Foundations of the Atomic Theory,* 1899 (Edinburgh: Alembic Club #2).

*——, 1808, *A new system of chemical philosophy;* 1842, Part I, 2nd ed. (London: Weale).

*Dampier, W.C., and Dampier, M., 1959, *Readings in the Literature of Science* (New York: Harper).

*Davy, H., 1839–1840, *The collected works of Sir Humphry Davy* (London: Smith, Elder & Co.); in Siegfried.

*——, *The elementary nature of chlorine* (including Bakerian Lectures of 1808, 1809, 1810); 1929 (London: Alembic Club #9).

——, *Decomposition of the fixed alkalies* (including Bakerian Lecture of 1807); 1901 (London: Alembic Club #6).

*de Broglie, L., 1923, "Waves and Quanta," *Comptes Rendus, 177.*

*——, 1923, "Light Quanta, Diffraction and Interference," *Comptes Rendus, 177.*

*——, 1924, "On the General Definition of the Correspondence between Waves and Motion," *Comptes Rendus, 179.*

*Descartes, R., 1637, *Discours de la Méthode Suivi des Méditations;* 1962 (Paris: Editions 10/18).

*——, *Philosophical Works,* 1955 (Haldane, E.S., and Ross, G.R.T., transl.) (New York: Dover).

Dictionary of National Biography, 1896 (London: Oxford Univ.).

*Donnay, J.D.H., and Harker, D., 1937, "A New Law of Crystal Morphology Extending the Law of Bravais," *American Mineralogist, 22,* p. 446.

Duhem, Pierre, 1905, *The aim and structure of physical theory;* 1954 (Wiener, P.P., transl.) (Princeton: Princeton Univ.).

*Dumas, J.B.A., 1834, "Recherches de chimie organique," *Ann. Chim. Phys., 56,* pp. 113–154; in Leicester and Klickstein.

*———, 1839, "Mémoire sur la constitution de quelques corps organiques et sur la théorie des substitutions," *Liebig, Annal., 32,* pp. 101–119.

*———, 1840, "Note sur la constitution de l'acide acétique et de l'acide chloracétique," *ibid., 33,* pp. 179–182.

*———, and Boullay, P., 1828, "Mémoire sur les éthers composés," *Ann. Chim. Phys., 37,* pp. 15–53; in Leicester and Klickstein.

*———, and Liebig, J., 1837, "Note sur l'état actuel de la chimie organique," *Comptes Rendus, 5,* pp. 567–572; in Leicester and Klickstein.

*Einhard, *The Life of Charlemagne,* 1960 (Turner, S.E., transl.) (Ann Arbor: Univ. of Michigan).

*Einstein, A., 1905, "Über Einen die Erzeugung und Verwandlung des Lichtes Betreffenden heuristischen Gesichtspunkt," *Ann. d. Physik, 17,* p. 132.

*———, 1906, "Zur Theorie der Lichterzeugung und Lichtabsorption," *Ann. d. Physik, 20,* p. 199.

*———, 1907, "Die Plancksche Theorie der Strahlung und die Theorie der Spezifischen Wärme," *Ann. d. Physik, 22,* p. 180.

*———, *Theory of the Brownian Movement,* 1956 (New York: Dover).

*Euclides, *Euclid's Elements;* in *Great Books of the Western World,* vol. 11, 1955 (Heath, T., transl.) (Chicago: Encyclopedia Britannica).

*Ewald, P. P., 1962, *Fifty Years of X-ray Diffraction* (Utrecht: International Union of Crystallography).

*Faraday, M., 1844, "Nature of Matter," *Phil. Mag., 24,* p. 136; in Guerlac, 1953.

*———, *Faraday's Diary,* 1932 (Martin, T., ed.), 8 vols. (London: Bell).

Farrington, B., 1949, *Greek Science* (Harmondsworth: Penguin).

Forbes, R.J., 1955–64, *Studies in Ancient Technology,* 8 vols. (Leiden: Brill).

*Galileo, 1638, *Two new sciences;* 1914 (Crew, H., and de Salvio, A., transl.) (New York: Macmillan).

*Galvani, L., 1791, extracts from *De Bononiensi Scientarum, 7;* in Magie.

*Gay-Lussac, L.J., 1809, "Sur la combinaison des substances gazeuses," *Mémoires de la Société d' Arcueil, ii,* p. 207 (Univ. of Chicago, Nat. Sciences 1, Selected Readings, 3rd ed.); in Guerlac, 1953.

*———, 1815, "Experiments on prussic acid," *Ann. Chim., 95,* pp. 136–231.

*Geiger, H., and Marsden, E., 1909, "On a diffuse reflection of the α-particles," *Proc. Roy. Soc., 82,* p. 495.

*Gerhardt, C., 1853, "Recherches sur les acides organiques anhydres," *Ann. Chim. Phys., 37,* pp. 285–342.

*Gibbs, J.W., 1906, *The Scientific Papers* (New York: Longmans, Green).

———, 1960, *Elementary Principles in Statistical Mechanics* (New York: Dover).

Giekie, A., 1905, *The Founders of Geology;* 1962 (New York: Dover).

Goodrich, L.C., 1963, *A Short History of the Chinese People* (New York: Harper).

*Guerlac, Henry, 1953, *Selected Readings in the History of Science,* vols. I and II (Ithaca: Cornell Univ. Library).

———, 1957, "Joseph Black and Fixed Air," I, II, *Isis, 48,* pp. 124, 443.

———, 1961, "Some Daltonian Doubts," *ibid.,* 52, p. 544.

*Hales, S., 1727, *Statical Essays: Containing Vegetable Staticks,* 4th ed. (London: Wilson & Nicol).

*Haüy, R.J., 1801, *Traité de Minéralogie* (Paris: Chez Louis).

Heisenberg, W., 1957, "Atoms with Hooks and Eyes," *Atlantic Monthly, 199,* 3, p. 121.

*Hertz, H., 1900, *Electric waves;* extracts in Guerlac, 1953 (Jones, D.E., transl.).

*Hittorf, J.W., 1869, "Über die Elektricitätsleitung der Gase," *Ann. d. Physik und Chemie, 136,* p. 1.

*Hofmann, A.W., 1851, "Researches into the Molecular Const. of the Organic Bases, 2nd Mem.," *Phil. Trans.,* pp. 357–398.

Holton, G., and Roller, D.H.D., 1958, *Foundations of Modern Physical Science* (Reading, Mass.: Addison-Wesley).

*Hooke, R., 1665, *Micrographia;* 1961 (London: Martyn and Allestry; New York: Dover).

*———, 1667, "Extract on Air, Respiration, and Lungs," *Roy. Soc. London, Phil. Trans.* 2, p. 539; in Guerlac, 1953.

*———, *The posthumous works,* 1705 (Waller, R., ed.) (London).

*Huygens, C., 1690, *Treatise on light;* 1962 (S.P. Thompson, transl.) (New York: Dover).

*Joule, J.P., 1887, *Scientific papers;* 1963 (London: Dawsons).

*Katchalski, E., Grossfeld, I., and Frankel, M., 1947, "Polylysine," *Jour. of the Amer. Chem. Soc., 69,* p. 2564.

*Kelvin, Lord (Thomson, W.), 1851, "On the Dynamical Theory of Heat, With Numerical Results Deduced From Mr. Joule's Equivalent of a Thermal Unit, and M. Regnault's Observations on Steam," *Trans. Roy. Soc. Edinburgh;* in *Scientific Memoirs, VI,* 1899 (Ames, J.S., ed.) (New York: Harper).

*———, 1882–1911, *Mathematical and Physical papers,* 6 vols. (Cambridge: Cambridge Univ.).

*Kepler, J., 1611, *Strena Seu de nive sexangula* (Frankfurt: Tampach).

*———, 1619, *Harmonices mundi,* V (Linz: Plank).

Kerker, M., 1955, "Hermann Boerhaave and the Development of Pneumatic Chemistry," *Isis, 46,* p. 36.

——, 1957, "Sadi Carnot," *Scientific Monthly, 85,* September, p. 3.

Laming, A., 1959, *Lascaux* (Baltimore: Penguin).

*Laplace, P.S., Marquis de, *Essay on probabilities,* 1951 (New York: Dover).

*von Laue, M., 1912, "Interferenz-Erscheinungen bei Röntgenstrahlen," *Sitzungsb, Bayer, Akad. Wiss., 42,* p. 303.

*Lavoisier, A.L., 1789, *Traité Élémentaire de Chimie;* in *Great Books of the Western World,* vol. 45, 1952 (Kerr, R., transl.) (Chicago: Encyclopedia Britannica).

*LeBel, J.A., 1874, "On the Relations Which Exist Between the Atomic Formulas of Organic Compounds and the Rotatory Power of Their Solutions," *Bull. Soc. Chim. de Paris, 24,* p. 337; in *Scientific Memoirs,* XIII, 1901 (Richardson, G.M., transl.) (New York: American Book).

*Leibniz, G.W., 1714, *Discourse on Metaphysics; . . . Monadology;* 1962 (G.R., transl.) (LaSalle, Ill.: Open Court).

*Leicester, H. M., and Klickstein, H.S., 1952, *A Source Book in Chemistry, 1400–1900* (New York: McGraw-Hill).

*Lémery, N., 1683, *Cours de Chymie* (Paris: author); extract in Guerlac, 1953.

*Leonardi, C., 1516, *Speculum Lapidum* (Venice: M. Sessa), 1717 (*Speculum Lapidum . . . Sympathia Septem Metallorum ac Septem Selectorum Lapidum ad Planetas*) (Hamburg: P. C. Albini).

*Leonardo da Vinci, *Notebooks,* 1935 (McCurdy, E., ed. and transl.) (New York: Empire).

Levene, P.A., and Bass, L.W., 1931, *Nucleic Acids* (New York: A.C.S. Mono).

Lewis, B., 1960, *The Arabs in History* (New York: Harper).

*Loschmidt, J., 1866, "On the Size of the Air-Molecule," *Chemisches Centralblatt, 11,* p. 31; reprint from *Sitz. d. Wiener Akad. der Wissen., Bd. 52,* p. 395, 1865.

*Lower, R., 1669, "A Treatise on the Heart, on the Movement and Colour of the Blood" (London); extract in Guerlac, 1953.

*Magie, W.F., 1935, *A Source Book in Physics* (New York: McGraw-Hill).

Manes, G.I., 1956, "The Discovery of X-ray," *Isis, 47,* p. 236.

Margenau, H., 1950, *The Nature of Physical Reality* (New York: McGraw-Hill).

——, and Lindsay, R.B., 1957, *Foundations of Physics* (New York: Dover).

Mason, S.F., 1962, *A History of the Sciences* (New York: Collier).

*Maxwell, J.C., 1860, "Illustrations of the Dynamical Theory of Gases," *Phil. Mag.* 4th ser., *19;* extracts in Magie.

*——, 1873, *Matter and motion;* 1952 (New York: Dover).

*Mayow, J., 1674, "Medico-physical Works"; 1908 (Edinburgh); extract in Guerlac, 1953.

Metzger, Hélène, 1918, *La Genèse de la science des cristaux, thèse* (Paris: Alcan).

*Michelson, A.A., and Morley, E.W., 1887, "The Michelson-Morley Experiment," *Phil. Mag., 5;* extract in Magie.

Middleton, W.E. Knowles, 1963, "The Place of Torricelli in the History of the Barometer," *Isis, 54,* p. 11.

*Milton, J., 1644, *Areopagitica* and *Of Education;* 1951 (New York: Appleton).

Moulton, F.R., and Schifferes, J.J., 1945, *The Autobiography of Science* (Garden City: Doubleday).

*Muirhead, J.P., 1854, "A Letter of James Watt to M. de Luc on the Trials of Joseph Priestley," in *The Origin and Progress of the Mechanical Inventions of James Watt,* vol. II (London); in Guerlac, 1953.

Nash, Leonard K., 1950, *The Atomic-Molecular Theory* (Cambridge, Mass.: Harvard Univ.).

——, 1956, "The Origin of Dalton's Atomic Theory," *Isis, 47,* p. 101.

Needham, J., 1954–1962, *Science and Civilization in China,* vols. 1–4 (Cambridge, England: Cambridge Univ.).

*Newlands, J.A.R., 1863, "On Relations Among the Equivalents," *Chemical News, vii,* p. 70.

*———, 1864, "Relations Between Equivalents," *ibid., x,* pp. 59, 94, 240.

*———, 1865, "On the Law of Octaves," *ibid., xii,* p. 83.

*———, 1865, "On the Cause of Numerical Relationships Among the Equivalents," *ibid.,* p. 94.

*———, 1866, "On the 'Law of Octaves,'" *ibid., xiii,* pp. 113, 130.

*Newton, I., 1686, 1729 (Motte, A., transl.), *Mathematical Principles of National Philosophy* (*Principia*); 1962 (Berkeley: Univ. of Calif.).

*———, 1704, *Opticks;* 1952 (New York: Dover).

*Odling, W., 1857, "On the Natural Groupings of the Elements," *Phil. Mag.,* 4th ser., *13,* p. 429.

Oparin, A., and Fesenkov, V., 1960, *The Universe* (Moscow: Foreign Languages Publishing House).

Ostwald W., 1910, *Natural philosophy* (New York: Holt).

*———, and Oettingen, A.J.V., 1887–1893, 1894, *Klassiker der Exakten Wissenschaften* (Leipzig: Engelmann).

*Pacioli, Luca, 1509, *De Divina Proportione;* 1959 (Span. transl. by Resta, R.) (Buenos Aires: Losada).

Partington, J.R., 1960, *A Short History of Chemistry* (New York: Harper).

———, 1961, *A History of Chemistry* (New York: St. Martin's).

*Pasteur, L., 1860, "Researches on the Molecular Asymmetry of Natural Organic Products"; 1905 (A.S., ed.) (Edinburgh: Alembic Club #14, W.F. Clay).

*Pauling, L., 1929, "The Principles Determining the Structure of Complex Ionic Crystals," *Jour. of the Amer. Chem. Soc., 51,* p. 1010.

*———, 1960, *The Nature of the Chemical Bond* (Ithaca: Cornell Univ.).

*Perrin, J., 1895, "Nouvelles Propriétés des Rayons Cathodiques," *Comptes Rendus, 121;* extract in Magie.

——, 1923 (Hammick, D., transl.), *Atoms* (London: Constable); extract in Guerlac, 1953.

*Perutz, M.F., 1964, "The Hemoglobin Molecule," *Scientific American, 211,* 5, p. 64.

*Petit, A.T., and Dulong, P.L., 1819, "Théorie de la Chaleur," *Ann. Chim. Phys., X,* p. 395; *Phil. Mag., 54,* p. 267.

Pirenne, H., 1958, *A History of Europe* (New York: Doubleday).

*Planck, M., 1901, "On the Law of Distribution of Energy in the Normal Spectrum," *Ann. d. Physik, 4,* p. 553.

*——, 1926, *Thermodynamics;* 1960 (New York: Dover).

*Plato, "Timaeus, Parmenides, Cratylus"; in *Dialogues,* 1953 (Jowett, B., transl.) (Oxford: Clarendon).

*Pope, W.J., and Barlow, W., 1907, "The Relation Between the Crystalline Form and the Chemical Constitution of Simple Inorganic Substances," *Chem. Soc. of London, Trans. 91,* II, p. 1150.

*Priestley, J., 1774, *The discovery of oxygen;* 1912 (Chicago: Alembic Club #7).

*——, *Scientific Correspondence,* 1892 (Bolton, H.C., ed.) (New York: private).

*——, *Considerations on the Doctrine of Phlogiston, and the Decomposition of Water,* 1929 (Foster, W., ed.) (Princeton: Princeton Univ.); extracts in Guerlac, 1953.

*Proust, J.L., 1799, *Ann. Chim., 32,* pp. 26–54; in Leicester and Klickstein.

*Prout, W., 1815, "On the Relation Between the Specific Gravities of Bodies in Their Gaseous State and the Weights of Their Atoms," *Ann. of Phil., 6,* p. 321; 1932 (Edinburgh: Alembic Club #20).

*Rey, J., 1630, *Essays of Jean Rey* (The increase in weight of tin and lead on calcination); 1904 (Edinburgh: Alembic Club #11).

*Richter, J.B., 1792–1794, *Anfangsgründe der stöchyometrie oder messkunst chymischer elemente,* 3 vols. (Breslau: Hirschberg).

*Roentgen, W.C., 1895–1898, "On a New Kind of Ray," *Ann. d. Physik und Chemie, 64;* in *Scientific Memoirs* (Ames, J.S., ed.) (New York: American Book).

Romer, A., 1960, *The Restless Atom* (Garden City: Anchor).

*Rumford, Count (Thompson, B.), 1870–1875, *The Complete Works of Count Rumford* (Boston: Am. Acad. Arts and Sciences).

*Rutherford, E., 1911, "The Scattering of α- and β-Particles by Matter and the Structure of the Atom," *Phil. Mag., XXI,* p. 669.

Sahyun, M., 1944, *Outline of the Amino Acids and Proteins* (New York: Reinhold).

*Sanger, F., 1945, "The Free Amino Groups of Insulin," *Biochem. Jour., 39,* p. 507.

*———, and Tuppy, H., 1951, "The Amino-acid Sequence in the Phenylalanine Chain of Insulin," *Biochem. Jour., 49,* p. 463.

*———, and Smith, L.F., 1957, "The Structure of Insulin," *Endeavour, 16,* p. 48.

*———, Thompson, E.O.P., and Tuppy, H., 1952, "Amino Acid Sequences in Insulin," *2nd Congr. Intern. Biochim., Chim., Biol., IV, Symposium Hormones Proteines et Derivées des Proteines,* p. 26.

Sarton, G., 1952, *A History of Science* (Cambridge, Mass.: Harvard Univ.).

*Scheele, C.W., 1774; Berthollet, C.L., 1785; de Morveau, G., 1787; Gay-Lussac, J.L., and Thénard, L.J., 1809; in *The Early History of Chlorine,* 1905, 1944 (Edinburgh: Alembic Club #13).

Schmidt, C.L.A., 1938, *Chemistry of the Amino Acids and Proteins* (Baltimore: Thomas).

Schneer, C.J., 1960, "Kepler's New Year's Gift of a Snowflake," *Isis, 51,* 4, p. 531.

*———, 1968, "Crystal Form and Crystal Structure," *Helvetica Physica Acta, 41,* p. 1151.

*Schrödinger, E., 1926, "An Undulatory Theory of the Mechanics of Atoms and Molecules," *Phys. Rev., 28,* p. 1049.

*Shamos, M.H., 1959, *Great Experiments in Physics* (New York: Holt).

Siegfried, R., 1963, "The Discovery of Potassium and Sodium," *Isis, 54,* p. 247.

Singer, C., Holmyard, E.J., and Hall, A.R., 1954–1958, *A History of Technology,* vols. I–V (New York: Oxford).

Smith, T.V., 1956, *From Thales to Plato* (Chicago: Univ. of Chicago).

*Sprat, R., 1667, *History of the Royal Society;* 1958 (London: Martyn [facs.]; St. Louis: Washington Univ.).

*Stahl, G., 1730 (Shaw, P., transl.), *Philosophical principles of universal chemistry* (London); extracts in Guerlac, 1953.

*Steno, Nicolaus, 1669, *De solido intra solidum naturaliter contento dissertationis prodromus* (Florence: Stella); 1671 (Henry Oldenburg, transl.) (London: Winter); 1916 (Winter, J.G., transl.) (New York: Macmillan).

Stillman, J.M., 1960, *The Story of Alchemy and Early Chemistry* (New York: Dover).

Stimson, D., 1948, *Scientists and Amateurs* (New York: Schuman).

Thomson, J.J., 1937, *Recollections and reflections* (New York: Macmillan).

*———, 1937, *Discovery of the Electron;* in Guerlac, 1953.

*Thomson, T., 1830–31, *History of chemistry* (London: Colburn & Bentley).

Thorpe, T.E., 1906, *Joseph Priestley* (New York: Dutton).

———, 1931, *Essays in Historical Chemistry,* 3rd ed. (New York: Macmillan).

*Torricelli, Evangelista, 1644, *The Barometer;* extract in Magie.

*van Helmont, J.B., 1662, *Oriatrike, or Physik Refined* (London); extract in Guerlac, 1953.

van Melsen, A.G., 1952, *From Atomos to Atom* (Pittsburgh: Duquesne Univ.).

*van't Hoff, J. H., 1874, "A Suggestion Looking to the Extension into Space of the Structural Formulas at Present Used in Chemistry" (Utrecht); in *Scientific Memoirs, XIII,* 1901 (Richardson, G.M., transl.) (New York: American Book).

*du Vigneaud, V., Ressler, C., Swan, J.M., Roberts, C.W., Katsoyannis, P.G., Gordon, S., 1953, "The Synthesis of an Octapeptide Amide with the Hormonal Activity of Oxytocin," *Jour. of the Amer. Chem. Soc., 75,* p. 4879.

*Volta, A., 1800, "On the electricity excited by the mere contact of conducting substances of different kinds," *Phil. Trans.,* pp. 403–431; in Magie.

*Wald, G., 1954, "The Origin of Life," *Scientific American, 191,* pp. 2, 4.

*Watt, J., *Correspondence of the Late James Watt on His Discovery of the Theory of the Composition of Water,* 1846 (Muirhead, J.P., ed.) (London: Murray).

Weyl, Hermann, 1952, *Symmetry* (Princeton: Princeton Univ.).

Whittaker, E., 1960, A History of the Theories of Aether and Electricity (New York: Harper).

*Williamson, A. W., 1850–56, *Papers on Etherification and on the Constitution of Salts;* 1949 (Edinburgh: Alembic Club #16).

*Wöhler, F., 1828, "Kunstl. Bildung d. Harnstoffs," *Poggendorff's Ann. der Physik, XII; Ann. Chim., 37,* pp. 330–334.

———, and Liebig, J., 1832, "Researches respecting the radical of benzoic acid," *Ann. Chimie, 3,* pp. 249–282; in Leicester and Klickstein.

Wolf, A., 1959, *A History of Science, Technology and Philosophy in the Sixteenth and Seventeenth Centuries* (New York: Harper).

———, 1961, *A History of Science, Technology and Philosophy in the Eighteenth Century* (New York: Harper).

*Woodward, R.B., and Schramm, C.H., 1947, "Synthesis of Protein Analogs," *Jour. of the Amer. Chem. Soc.*, *69*, pp. 1551–1552.
*Wurtz, C.A., 1949, "Sur une série d'alcalis organiques homologues avec l'ammoniaque," *Comptes Rendus*, *28*, pp. 223–226.

Index